T0293230

Cybersecurity and Supply Chain Risk Management Are Not Simply Additive

Implications for Directions in Risk Assessment, Risk Mitigation, and Research to Secure the Supply of Defense Industrial Products

VICTORIA A. GREENFIELD, JONATHAN W. WELBURN, KAREN SCHWINDT, DANIEL ISH, ANDREW J. LOHN, GAVIN S. HARTNETT

Prepared for the Department of the Air Force
Approved for public release; distribution is unlimited.

 PROJECT AIR FORCE

For more information on this publication, visit **www.rand.org/t/RRA532-1**.

About RAND

The RAND Corporation is a research organization that develops solutions to public policy challenges to help make communities throughout the world safer and more secure, healthier and more prosperous. RAND is nonprofit, nonpartisan, and committed to the public interest. To learn more about RAND, visit www.rand.org.

Research Integrity

Our mission to help improve policy and decisionmaking through research and analysis is enabled through our core values of quality and objectivity and our unwavering commitment to the highest level of integrity and ethical behavior. To help ensure our research and analysis are rigorous, objective, and nonpartisan, we subject our research publications to a robust and exacting quality-assurance process; avoid both the appearance and reality of financial and other conflicts of interest through staff training, project screening, and a policy of mandatory disclosure; and pursue transparency in our research engagements through our commitment to the open publication of our research findings and recommendations, disclosure of the source of funding of published research, and policies to ensure intellectual independence. For more information, visit www.rand.org/about/research-integrity.

RAND's publications do not necessarily reflect the opinions of its research clients and sponsors.

Published by the RAND Corporation, Santa Monica, Calif.
© 2023 RAND Corporation
RAND® is a registered trademark.

Library of Congress Cataloging-in-Publication Data is available for this publication.

ISBN: 978-1-9774-1273-7

Cover: Adapted from Mycola/Getty Images and halbergman/Getty Images.

About This Report

In mid-2019, the Air Force Research Laboratory (AFRL) asked RAND Project AIR FORCE (PAF) for assistance understanding how cyber-related risks compare with other risks to its defense-industrial supply chains—a scope that included supply chains for hardware, not supply chains for software per se—and exploring implications for risk assessment and mitigation and for research. Over the next 18 months, PAF sought to characterize cyber-related risks to supply chains and identify directions for addressing the distinct—unique, exceptional, and sometimes-reinforcing—challenges that cyber-related risks pose to defense-industrial supply chains and, hence, to supply chain risk management (SCRM).

This report discusses that PAF research effort. The effort was part of a larger undertaking that also explored national security policies at the nexus of cybersecurity and SCRM, as well as tools and frameworks for addressing cyber-related risks. The report complements a body of recent RAND work, including several studies on the cybersecurity of Department of the Air Force weapon systems and industrial control systems, cyber vulnerabilities, and global supply chain risks. It should be of interest to those seeking to secure the supply of defense industrial products from the risks of cyberattacks, primarily from the perspective of SCRM, and across research and policy communities.

The research reported here was commissioned by AFRL and conducted within the Resource Management Program of RAND Project AIR FORCE as part of a fiscal year 2020 project, "Cybersecurity of the Air Force Industrial Supply Chain."

RAND Project AIR FORCE

RAND Project AIR FORCE (PAF), a division of the RAND Corporation, is the Department of the Air Force's (DAF's) federally funded research and development center for studies and analyses, supporting both the United States Air Force and the United States Space Force. PAF provides the DAF with independent analyses of policy alternatives affecting the development, employment, combat readiness, and support of current and future air, space, and cyber forces. Research is conducted in four programs: Strategy and Doctrine; Force Modernization and Employment; Resource Management; and Workforce, Development, and Health. The research reported here was prepared under contract FA7014-16-D-1000.

Additional information about PAF is available on our website:
www.rand.org/paf/

This report documents work originally shared with the DAF on December 30, 2020. The draft report, dated December 2020, was reviewed by formal peer reviewers and DAF subject-matter experts.

Acknowledgments

We are grateful to Timothy Sakulich, Director of the Materials and Manufacturing Directorate at the Air Force Research Laboratory (AFRL/RX) at Wright-Patterson Air Force Base, who sponsored this project and provided thoughtful and encouraging support throughout its duration. We are also grateful to Charles H. Ward, Manufacturing and Industrial Technologies Division Chief; Brenchley L. Boden II; Col Charles D. Ormsby; and the staff at AFRL/RX for their support, insightful conversation, and productive engagement throughout this project. In addition, we thank the authors of several prominent publications who shared their perspectives with us.

We are also grateful to our colleagues at RAND. We thank Justin Grana, Kelly Klima, Krista Langeland, Lindsay Polley, Adrienne Propp, and Jonathan Wong for their role in early conceptualizations of this project. We thank R. J. Briggs for his insight on private-sector incentives; Anita Szafran for her thorough assistance in our searches across different literatures; and Jared Ettinger, Caolionn O'Connell, and Don Snyder for their helpful advice and conversations. Don Snyder, along with Lloyd Dixon and Elizabeth Bodine-Baron, also provided valuable formal reviews of our draft report. Additionally, we are thankful to Rosa Maria Torres and Olatunda Martin for their administrative support and to Phyllis Gilmore for her outstanding editing of our work.

We are grateful to Stephanie Young, director of the Resource Management Program, and Obaid Younossi, former program director, for their useful advice at various stages of this project.

Any remaining errors or omissions are solely our responsibility.

Summary

Issue

Our analysis in this report and such events as a 2017 cyberattack that impaired commercial distribution globally and the 2020 SolarWinds breach lend credence to the view that costly cyberattacks have become an eventuality for many organizations. Against that backdrop, the Air Force Research Laboratory (AFRL) asked RAND Project AIR FORCE (PAF) for assistance understanding how cyber-related risks compare with other risks to its defense-industrial supply chains—a scope that included supply chains for hardware, not supply chains for software—and exploring implications for directions in risk assessment and mitigation and for research. AFRL was interested in how attackers might use supply chains to wage attacks, such as through malicious code, and how supply chains might, themselves, be targets of attack, such as through disruption.

Approach

Over an 18-month period, beginning in mid-2019, PAF sought to answer two research questions, "How do cyber-related risks differ from or compound other concerns about supply chain risk management (SCRM)?" and "What do, or could, these differences mean for risk assessment, risk mitigation, and research?" To conduct the analysis, PAF drew insights from the literatures on cybersecurity, SCRM, game theory, and network analysis and worked with sets of stylized supply chains and fundamental principles of risk management. The report uses the phrase *cyber SCRM* broadly, to refer to the cybersecurity of supply chains, including attacks *through* supply chains to reach a target and attacks *on* supply chains in which the target of the attack is the supply chain itself.

Key Insights

The first insight pertains directly to the first research question, and the subsequent insights pertain largely to the second research question, but with some overlap:

- **Cyber-related risks could be substantially worse than and different from others.** Cyber events can present the worst of all the characteristics of conventional hazards, judged in terms of their onset, duration, visibility, and reach, and can pose even greater challenges than nondigital threats, given the potential for strategic adversaries to inflict harm at low cost and without punishment of repeated attempts.
- **Preventative measures are not enough**. Preventative measures cannot stand alone or be pursued at the expense of taking steps to facilitate response and recovery or build

resilience. Creating impenetrable defenses is infeasible, and attempting to create them would entail further risks and costs.

- **Cyber SCRM requires more than an amalgam of *cyber* and *SCRM*.** The risks of cyberattacks might be unhindered or even elevated by some conventional means of addressing cyber and SCRM concerns—such as those for exploitation and disruption—separately, suggesting the potential for trade-offs among risk-reduction objectives. Absent any trade-offs, a fusion of cyber- and SCRM-based measures could be inadequate if conventional SCRM underestimates the potency of cyberattacks relative to other sources of risk.
- **Private-sector efforts to manage risk may not meet national security needs.** Strategic interactions between suppliers and attackers could lead to underinvestment in security, especially without coordination among suppliers, but compounding factors, involving risk assessment, incentives, and supply chain visibility, could make matters worse. A supplier that cannot see how far its supply chain reaches or the dependencies within it cannot be expected to mitigate risk to its own satisfaction, let alone to that of the Department of the Air Force (DAF).
- **Research can do more to support cyber SCRM.** This could include delving into the details of some long-standing issues, such as those regarding risk assessment, possibly with new or different analytical methods, and by exploring other issues that came to the fore in this research, such as those concerning private-sector engagement.

Directions for Cyber SCRM and Research

We suggest taking a comprehensive approach to cyber SCRM that would address cybersecurity and SCRM together by

- framing the potential consequences of cyberattacks in terms of the availability, quality, and cost of defense industrial products that serve mission-critical roles, not just or primarily in terms of information security
- establishing priorities among those cyber and SCRM consequences based on what they could mean for mission attainment
- setting out terms for cyber SCRM strategies, with due attention to response, recovery, and resilience, that account for concerns about
  - information security and supply chain functionality
  - differences in DAF and private-sector interests that could affect whether and how industry contributes to risk reduction
  - trade-offs among risk-reduction objectives, relating, for example, to supply chain disruption, on the one hand, and information vulnerability, on the other.

We also discourage dwelling too much on efforts to fine-tune risk assessments and build impenetrable defenses, which are unlikely to succeed and could distract from other risk reducing activities. Finally, we highlight opportunities for research in four areas that could strengthen the foundation for risk management:

- approaching risk assessment with realistic expectations and with greater emphasis on supply chain functionality
- establishing needs and priorities for responding to, recovering from, and increasing the resilience of supply chains to cyberattacks, especially in relation to supply chain functionality and mission attainment
- examining the utility and limits of private-sector risk reduction
- crafting a comprehensive strategy for cyber SCRM.

Contents

Figures and Tables

Figures

Tables

Chapter 1. Introduction

That an organization will experience a costly cyberattack tends to be discussed as a "when," not an "if."[1] For supply chains, for which third-party cybersecurity has become a prominent concern, describing such attacks as eventualities might seem reasonable. According to a survey conducted by the cybersecurity company BlueVoyant, 92 percent of respondents reported suffering a cyber intrusion in 2019–2020 as a direct result of third-party cybersecurity weakness in the supply chain (BlueVoyant, 2020). A 2017 attack, dubbed *NotPetya*, originated from an exploited vulnerability in a small accounting software company yet led to massive disruptions across Europe and resulted in costs in the millions of dollars for recovery, network sanitization, and lost sales (Nash, Castellanos, and Janofsky, 2018). That and other recent events, including a 2020 breach commonly referred to as *SolarWinds*—in which cyber intruders appear to have accessed and exploited software update mechanisms that would ordinarily contribute to security across government and industry—lend credence to perceptions of inevitability.[2]

Against that backdrop, in mid-2019, the Air Force Research Laboratory (AFRL) asked RAND Project AIR FORCE (PAF) to help it understand how cyber-related risks differ from or reinforce other risks to its defense-industrial supply chains—a scope that included supply chains for hardware, not supply chains for software—and to explore implications for risk assessment and mitigation and for research. AFRL's interest was not just in how "attackers" might use supply chains to wage attacks (e.g., through malicious insertion of code) but also in how supply chains might, themselves, constitute targets of attack (e.g., to cause disruption). (Box 1.1 discusses our use of *cyberattack*, and Appendix A discusses that and other prominent vocabulary in this report.) The difference being one of attackers acting *through* supply chains to reach a target versus acting *on* supply chains as targets. Over the next 18 months, PAF sought to characterize the risks of cyberattacks both through and on supply chains and to identify directions for risk assessment and mitigation and for research to address the distinct—unique, exceptional, and sometimes-reinforcing—challenges that cyber-related risks pose to those supply chains and, hence, to supply chain risk management (SCRM).[3]

This report discusses PAF's research effort. The effort was part of a larger undertaking that also explored national security policies at the nexus of cybersecurity and SCRM, as well as tools

---

[1] For examples of these types of statements, see Aldasoro et al. (2020) and Bartock et al. (2016).

[2] For an ongoing record of significant cyber episodes since 2006, see Center for Strategic and International Studies (undated). Although we have noted the occurrence of the 2020 SolarWinds attack, our analysis concluded before that attack occurred and, thus, does not cover it.

[3] We acknowledge that the policy community has not stood still since we completed the research for this report, as evident in new directives, guidance, and analysis. For examples of each, see Executive Order (EO) 14017 (2021), EO 14028 (2021), Boyens et al. (2015), U.S. Government Accountability Office (2020), and White House (2021).

and frameworks for addressing cyber-related risks. The report complements a body of prior RAND Corporation work, including several studies on the cybersecurity of Department of the Air Force (DAF) weapon systems and industrial control systems, cyber vulnerabilities, and global supply chain risks (documented in, e.g., Snyder et al., 2015; O'Connell et al., 2021; and Snyder et al., 2020). As we discuss in later chapters, the report also builds on other work on cybersecurity, SCRM, and related methods of analysis, including several governmental and nongovernmental reports (see, e.g., Defense Science Board, 2017; U.S. Department of Defense [DoD], 2018; National Defense Industrial Association [NDIA], 2017; Nissen et al., 2018; and Cyberspace Solarium Commission, 2020). Past work has tended to treat cybersecurity and SCRM as matters of information and supply surety, respectively, overlapping primarily in relation to *cyber* supply chains.[4] In the absence of perfect terminology for our purpose, our report uses the phrase *cyber SCRM* broadly, to refer to the cybersecurity of supply chains, writ large, taken to include attacks on supply chains in which the target of the attack is the supply chain itself.

Box 1.1. What Do We Mean by *Cyberattack*?

In many DoD contexts, *attack* carries a precise definition: "Actions taken in cyberspace that create noticeable denial effects (i.e., degradation, disruption, or destruction) in cyberspace or manipulation that leads to denial that appears in a physical domain, and is considered a form of fires" (Joint Publication 3-12, 2018, p. GL-4). A cyberattack, thus, would differ, for example, from an act of cyber exfiltration that extracts information and leads to a loss of confidentiality. However, across audiences and methods, including those in the disciplines that we used to construct our analysis, the term *cyberattack* is often used differently or more broadly. The National Institute of Standards and Technology (NIST), a widely referenced source of cybersecurity standards and definitions, offers the following definition for *cyberattack*: "An attack, via cyberspace, targeting an enterprise's use of cyberspace for the purpose of disrupting, disabling, destroying, or maliciously controlling a computing environment/infrastructure; or destroying the integrity of the data or stealing controlled information" (NIST Computer Security Resource Center, undated).[a]

In turn, NIST Computer Security Resource Center (undated) defines *attack* variously as, for example, "Any kind of malicious activity that attempts to collect, disrupt, deny, degrade, or destroy information system resources or the information itself" and as "The realization of some specific threat that impacts the confidentiality, integrity, accountability, or availability of a computational resource," and in terms of unauthorized entities' deceitful practices.

In this report, we adopt a broad perspective on attacks and attacking in our use of the term *cyberattack* and further specify types of actions or intrusions by their impact on the supply chain, highlighting both disruption and exploitation. We also work with the long-used terms in game theory of *attackers* and *defenders*, referring to agents in adversarial relationships. By implication, we are not using the terms *cyberattack* or *attack* to impart any legal weight or operational authority.

[a] See the NIST glossary entries for *cyberattack* and *attack* (NIST Computer Security Resource Center, undated).

---

[4] As discussed in Chapter 2, this is a broad generalization subject to important exceptions.

Research Aim and Questions

In light of AFRL's research request, this report attempts to characterize the risks of cyberattacks through and on supply chains both on their own and in relation to the risks of other malicious threats, such as bombings, arson, and product tampering, and hazards, such as floods, earthquakes, wildfires, mechanical or technical failures, and human errors. Our aim is to better understand how cyber-related risks—including the underlying threat environment—compare with other risks to defense industrial supply chains and to identify potential needs for new or different approaches to addressing the risks. We have framed our aim with two closely related research questions: "How do cyber-related risks differ from or compound other concerns about SCRM?" and "What do, or could, these differences mean for risk assessment, risk mitigation, and research?" To the extent that the risks of cyberattacks are like or unlike those of other, long-standing hazards and threats, attention to the differences could help shed light on the degree and form of attention cyberattacks require.

Figure 1.1 categorizes SCRM concerns broadly, in terms of risks of disruption and risks of exploitation, from either infiltration or exfiltration, and it shows where cyberattacks stand in relation to each.

Figure 1.1. Cyberattacks in Relation to Other SCRM Concerns

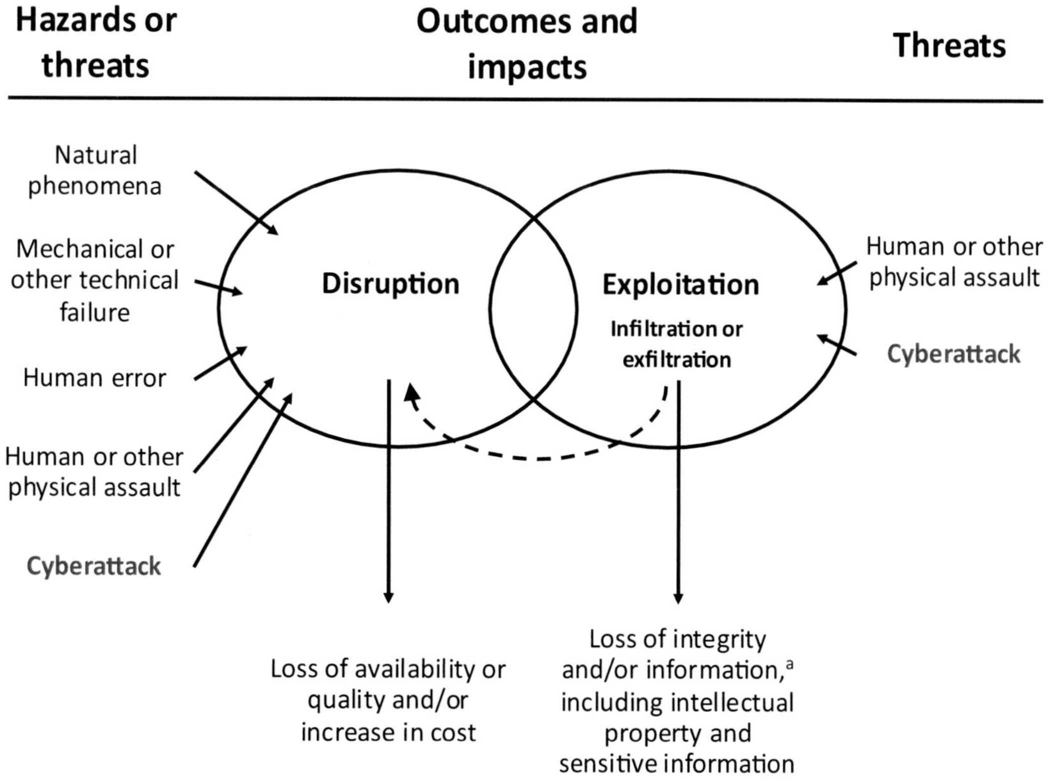

NOTE: The lists of hazard and threats are illustrative, not exhaustive.
[a] Information can also leak unintentionally through mechanical or technical failure or human error.

Threats and hazards of all types can affect the availability, quality, and cost of products through disruption.[5] Threats can also yield *exploitation*, by which we mean information-based effects. Infiltration might affect all three of these product dimensions—availability, quality, and cost—in the near or longer term. Exfiltration—or espionage—can play out differently, possibly with more intervening steps, over time. In a case of infiltration, an attacker might insert a malicious code that could affect—degrade or alter—the quality of a product directly, through a near or longer-term operational effect. In a case of exploitation, the effects could emerge less directly. For example, the exfiltration could entail a loss of information that enables counterfeiting or duplication, reducing the product's effectiveness and, hence, its quality. In either case, one might say the attack affected the integrity of the product, but the path of effects, as well as the effects themselves, could differ substantially between the cases. An adversary can also use exfiltrated information to identify weaknesses in products and plan a future attack.

One could also distinguish among risks to manufacturing (production or assembly) and logistics (moving or storing products, etc.). In a stylized visual representation of a supply chain, such as Figure 1.2, manufacturing would occur at the circular nodes,[6] which represent suppliers; transportation or distribution would occur along the solid arrows between nodes.[7] Although we did not call distributors out separately in our formal analyses, many of our findings, e.g., on interests, incentives, and redundancy, would apply to distributers as well. Information can also flow between and among suppliers, both across and within tiers.

In the remainder of this chapter, we discuss where our analytical approach resides in relation to real-world complexities, introduce our lines of effort, preview their roles in our analysis, and provide a road map to guide our readers.

The Real-World Complexities of Cyber-Related Risks

> *I have yet to see any problem, however complicated, which when looked at in the right way did not become still more complicated.*
> —*Poul Anderson*

Poul Anderson aptly sets out our—or any researcher's—analytical dilemma. Throughout this project, we routinely confronted complicated problems that, on further inspection, became much more complicated. The reality is that cyberattacks, as we define them (Box 1.1), are far from

---

[5] We use the term *availability* to cover quantity and timeliness, including scheduling, and *quality* to cover performance and any other issues that might entail a change in product characteristics.

[6] Throughout this report, we adopt the language of network analysis in which *node* (or, equivalently, *vertex* in graph theory) refers to a single entity connected to other entities in a network through *arcs* (or, equivalently, *edges* in graph theory).

[7] For stylized industrial supply chain depictions that incorporate additional features, including circular flows, internal sourcing among corporate divisions, alternative modes of distribution, and imports and exports, see, e.g., National Academies of Sciences, Engineering, and Medicine (2018).

Figure 1.2. Stylized Supply Chain

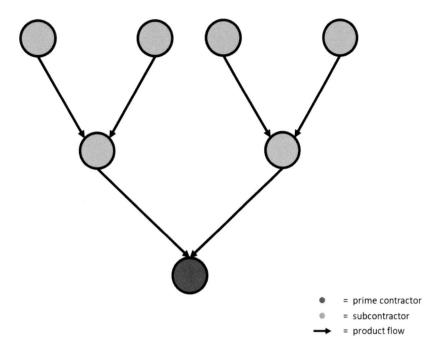

= prime contractor
= subcontractor
= product flow

simple. To achieve its ultimate objective, an attacker might compromise disparate systems in different ways—for example, by extracting information to gain knowledge from one target, using social engineering to gain system access to another target, and using a physics-based attack method (e.g., attacks that alter chips at the atomic level) to alter semiconductors deep within a supply chain to ultimately break an encryption—all before eventually delivering malware to a defender. Furthermore, the combination of a cyber realm and physical supply chains gives way to the possibility of network attacks on networks, which is a theme we often return to in this report. For those tasked with the challenge of defending the cybersecurity of supply chains, the number of attack modes and vectors can feel unbounded.

Another way of stating the Anderson quote is that, looked at the right way, a problem can become so complex that the complexity itself becomes debilitating. In this report, we attempt to navigate the complexity by leveraging various stylized models that capture the essence of one or more aspects of the problem—such as the juxtaposition of risk in different forms or the role of strategic behavior—and to discern common themes. In some, perhaps most, instances, our analyses are simple beyond tactical utility but can still shed light on the nature of cyber-related risks and on the implications for directions in risk assessment and mitigation and in research.

With that perspective in mind, we describe our lines of effort—the approaches and methods that we used in our analysis to answer our research questions in Chapter 2 and related appendixes—but are cognizant of their limitations.

Readers' Guide

In this project, we sought to understand how the risks of cyberattacks on defense-industrial supply chains—specifically, for hardware—differ from or reinforce the risks of other threats and hazards and the implications for directions in risk assessment and mitigation and in research. Thus, the report addresses how the peculiarities of cyber-related risks, including the threat environment, might interact with and affect defense-industrial supply chains and translate into distinct challenges for cyber SCRM. Cyber-related risks have many noteworthy attributes, but the importance for our research lies in the relevance for ensuring that DAF can get what it needs, when it needs it, at an acceptable cost.

On that basis, this report proceeds as follows:

In Chapter 2, we describe our lines of effort, consisting of the approaches and methods that we used in our analysis to answer our research questions.

In Chapters 3, 4, and 5, we consider attributes of cyber-related risks that pose noteworthy—unique, exceptional, or reinforcing—challenges to SCRM through the lenses of risk management, Boolean logic, game theory, and network analysis. In Chapter 3, we identify a range of commonly cited attributes, relating to the onset, duration, visibility, and reach of an attack, and compare them to various natural hazards, including floods, earthquakes, wildfires, and infectious disease. In Chapters 4 and 5, we consider, first, how behavior among attackers and defenders contributes to risk and, second, the role of supply chains themselves in propagating or amplifying risk. Collectively, these chapters should be of interest to readers who would like to know "what makes cyber-related risks different?" for defense industrial supply chains and for SCRM.

In these chapters, we employed different methods to characterize—and differentiate—cyber-related risks, but the methods also revealed needs for addressing risk and for further research, which we return to in Chapter 6. In Chapter 6, we summarize our key findings in relation to each research question, discuss the analytical insights that support these findings, and elaborate on our findings by considering possible directions for risk assessment and mitigation and for research. This chapter should be of interest to the policy community, industry, and researchers within and beyond DAF.

We provide technical appendixes that flesh out some of our terminology (Appendix A); provide more detail on opportunities for research on cyber SCRM (Appendix B); discuss foundations in game theory that bear on cyber SCRM (Appendix C); present and explore mathematical representations of our stylized supply chains along with related literature (Appendix D); and discuss the risk-management framework and Boolean attack model in greater detail, both on their own and in relation to NIST's cybersecurity framework (Appendix E).

Chapter 2. Lines of Effort

In this chapter, we describe the approaches and methods that we used in our analysis to answer each research question. Given the multidimensional nature of cyber SCRM, we employed multiple lines of effort to address our research questions and, relatedly, to explore the intersection of cybersecurity and SCRM challenges. These lines included building on and drawing insight from wide-ranging academic and nonacademic literatures, including those pertaining to SCRM, cybersecurity, game theory, and network analysis, and exploring risk transmission and outcomes in stylized supply chains. Throughout, we also drew from common military risk management methods, including a five-step risk management process and associated risk assessment matrix, as well as a Boolean attack model. In effect, we tugged on several analytical threads that span literatures, methods, and policy communities. Recalling our discussion of complexity in Chapter 1, we acknowledge that no single line of effort can capture all the nuances of cyber-related risks or how to address them but believe that these, taken together, provide insight into answer our research questions.

Figure 2.1 depicts our approach, although it belies substantial iteration.

Figure 2.1. Integration of Multiple Methods

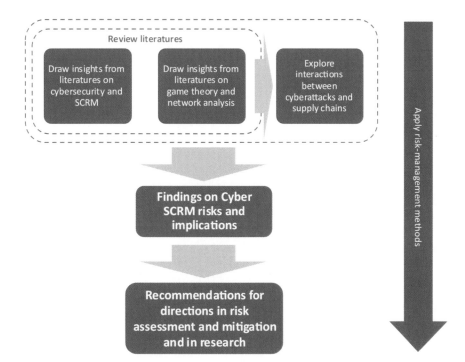

7

Literature Reviews

We looked to the literature to better understand how cyber-related risks to defense-industrial supply chains—specifically, hardware—differ from or reinforce other risks and the implications of any differences for risk assessment and mitigation.[8] That is, we sought to understand both the distinct challenges of the risks and the means of addressing the challenges. Although our object of interest was industrial supply, not cybersecurity per se, we could not examine their interface without considering both industrial supply and cybersecurity. That, in turn, also required considering underlying behaviors (i.e., those of attackers and defenders), supply chain configurations (e.g., networks), and various uncertainties. To span these domains topically and methodologically, we focused especially on four literatures: cybersecurity, SCRM, game theory, and network analysis. The first two literatures presented obvious avenues of pursuit, but concerns about behavior, supply chain configuration, and uncertainty led us to the third and fourth.[9]

We refer to the cybersecurity and SCRM literatures separately because we found less evidence of a distinct cyber SCRM literature than of potentially relevant strands of each literature and an amalgam of the two. The literature on cybersecurity, much as that on cybersecurity, most typically focuses on the security of information; the literature on SCRM, like its practice, tends to address the surety of supply. Sometimes, the literatures overlap, as when they consider flows of supply of information technology (IT)–related products, or merge together to form a more-fused literature on cyber SCRM as we define it. Along those lines, Boyson (2014, p. 342) recognizes cyber supply chain risk management as a discipline, but the author identifies this discipline as "resulting from the fusion of approaches, methods, and practices from the fields of cybersecurity, enterprise risk management, and supply chain management."

---

[8] We define *literature* broadly to encompass both academic and nonacademic sources, including publicly available peer-reviewed publications; conference proceedings; government, industry, and media reports; and commentary. An initial, informal exploration of academic work, focusing on methodological and taxonomical research, yielded 137 potentially relevant articles and conference papers. It also pointed us to related work, through citations and references, and suggested keywords and phrases to use in subsequent, more-formal searches. In addition, several ad hoc searches, e.g., using Google, Google Scholar, and the Defense Technical Information Center, yielded a few dozen or more articles, theses, conference papers, and press reports. Three more-formal efforts generated several hundred additional unique results. For example, the first such effort explored the nexus of supply-related and cyber-related terminology, using such search strings as "supply chain* AND cybersecurity" (("supply chain*" OR "defense industry" OR "defense industrial base" OR supplier*) AND (cybersecurity OR "cyber security" OR "cyber disrupt*" OR "cyber vulnerab*" OR "cyber risk*" OR "cyber breach*" OR "cyber attack*" OR "cyber threat*" OR "cyber incident*")), and yielded 502 unique results. Another effort replaced the cyber-related terminology with "IT" and other information-related terminology. In each of these cases, we limited the results to peer-reviewed work located with the Academic Search Complete, Association for Computing Machinery (commonly known as *ACM*), Business Source Complete, Institute of Electrical and Electronics Engineers (commonly known as *IEEE*), Scopus, and Web of Science search engines; limited terms to the title or abstract; used a truncation symbol ("*") to search word variations; and limited the date range to 2000–2020.

[9] We do not call out work on "risk assessment and mitigation" separately because it underpins both literatures and, arguably, much of the other literatures on our list. We return to this later.

We also found that the SCRM literature tends to take a private-sector perspective, which can limit its applicability to DAF's concerns or necessitate extrapolation and adaptation.[10] As we address later in this report, businesses typically strive to meet their objectives for profitability, rates of return, etc., while national security agencies, such as DAF, typically strive to meet public needs, as set out in laws, regulations, and policy. Still, a few recent governmental and nongovernmental reports have addressed concerns that we raise in this report (see, e.g., Defense Science Board, 2017; DoD, 2018; NDIA, 2017; Nissen et al., 2018; and the Cyberspace Solarium Commission, 2020).

Similarly, game theory models and network analysis sometimes address cyber SCRM directly, in ways that speak to DAF's concerns, but drawing insight from them often requires extrapolation and adaptation from other contexts. In a typical attacker-defender model, a defender might take some protective action (e.g., invest in enhanced security measures), and an attacker might respond to the action. The response might depend partly on whether the attacker behaves randomly (e.g., as in the case of a broad-based phishing attack) or strategically (e.g., by selecting among targets based on their perceived value or vulnerability) and on the attacker's objectives. Network models can be used to trace the effects of different types of risks on different types of supply chains, which can differ by configuration. In particular, the contours of the supply chain can serve to amplify or dampen the overall impact of a cyberattack and can affect who along the chain bears the brunt of it.

While it was impossible in the context of this research to delve fully into each literature, we drew important lessons and themes from each that we highlight chapter by chapter and in related appendixes (Appendixes B and C). For example, our initial exploration of the literature suggested the relevance of comparing cyber-related and other supply chain risks. It also warned us of the difficulty of categorizing, let alone assessing, risk and against focusing on preventing cyberattacks at the expense of response and recovery.[11] Regarding the latter, the Cyberspace Solarium Commission (2020), the Defense Science Board (2017), and Bartock et al. (2016) have suggested greater attention to response and recovery. As Bartock et al. (p. vi) observed, "There has been widespread recognition that some of these cybersecurity (cyber) events cannot be stopped and [that] solely focusing on preventing cyber events from occurring is a flawed approach."

---

[10] Work on SCRM is undergoing a rapid evolution and increase in interest, as reflected in an approximate doubling of the publication rate on "supply chain resilience" from the 2000s to the 2010s (Moore et al., 2015; Kamalahmadi and Parast, 2016). Consequently, the topic may have shifted in emphasis or further broadened its purview since the time of our research. Even at that time, the scope of the risks considered was, by its nature, significantly broader than that of the cybersecurity community, including demand uncertainty (see Baghalian, Rezapour, and Farahani, 2013), natural disasters, the political and regulatory environment (see Fiksel, 2015), and the emergent dynamics of supply networks (see Dolgui, Ivanov, and Sokolov, 2018).

[11] For more on *response, recovery*, and related concepts, see, e.g., NIST (2018a), NIST (2018b), and Appendix E, which discusses NIST (2018b).

We also reached out to the authors of some prominent studies to gain a fuller picture of their methods and findings. Their published work suggested tensions between firms' participation in defense-industrial supply chains and the costs of complying with cybersecurity requirements, differences among and between incentives in the private and public sectors, and differences between threats to IT and operating technology (OT), all of which could affect defense industrial supply chains and cyber SCRM (see, e.g., Simon and Omar, 2020; Defense Science Board, 2017; Nissen et al., 2018; and NDIA, 2017).

Stylized Supply Chains

Building on models of game theory and network analysis, we developed a set of highly stylized supply chains, much like that in Figure 1.2, to trace the flow of damage from each different type of attack—disruptive or exploitative—under different circumstances, with special attention to supply chain configuration. Our stylized renderings, coupled with simple mathematical representations of threats, as presented in Chapter 5 and Appendix D, enabled us to consider the implications of supply chain configuration—such as the number of tiers in a chain or the number of suppliers in a single tier—for disruption and exploitation and to identify potential trade-offs among risk-reduction objectives. Notwithstanding their simplicity, we can use the renderings to consider real-world questions, such as "what happens if supply expands?" or "what happens if suppliers share vulnerabilities?"

Methods for Managing Risk

Throughout this report, a small set of methods to support risk management, consisting of a long-standing risk management framework and a Boolean attack model (see Appendix E), influenced our thinking on cyber-related risks and SCRM. We drew on these methods most explicitly at the start, to characterize cyber-related risks and compare them with others, and at the finish, in our discussions of their implications for cyber SCRM. We chose to work with this risk management framework rather than the framework presented in Air Force Instruction 17-101 (2020) or the NIST cybersecurity framework (NIST, 2018b) because of its broader applicability and familiarity outside the cyber community. In Appendix E, we explore these methods in greater detail and map the risk management framework to the cybersecurity functions in NIST's cybersecurity framework.

The framework that we have adopted includes a five-step risk management process (Figure 2.2) that covers risk assessment and mitigation in wide-ranging military and other contexts and a risk matrix (Figure 2.3) that can be used to assess risk.[12] In this context, *risk mitigation*, which begins with "develop controls," can include measures that address response,

---

[12] This framework differs both in its steps and generality from NIST's cyber-specific framework (2018b), but, as we show in Appendix E, the steps are generally consistent with NIST's approach.

recovery, and resilience.[13] Developing controls does not equate with eliminating—or defending away—risk but can instead mean reducing risk to an acceptable "residual" (Greenfield and Camm 2005, p. 49; Air Force Pamphlet [AFPAM] 90-803, 2017, pp. 27–36; and Department of the Army Pamphlet 385-30, 2014, pp. 9–10). For example, a "control" could reduce a risk by making the outcome of an attack less onerous. In the five-step process, risk assessment informs risk mitigation, but its absence can also constrain mitigation. Looking for a pragmatic path to reconciling this tension, we consider opportunities for leveraging the sequential process iteratively and continuously in Chapter 6.

Figure 2.2. Five-Step Risk Management Process

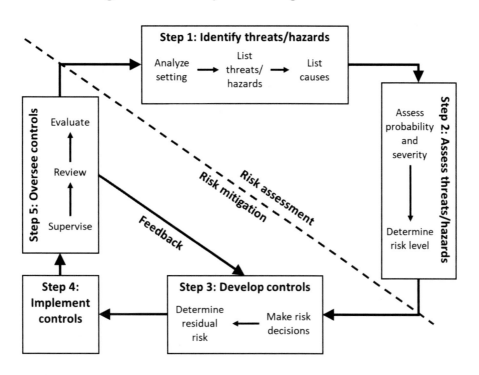

SOURCE: Authors' adaptation of Greenfield and Camm (2005), based on U.S. military guidance.

In the risk matrix, probability and severity jointly determine risk and can suggest priorities for mitigation (see Box 1.1 and Appendix A for vocabulary). We treat the matrix, which places probability on one axis and severity on the other, as a heuristic device in our analysis because it can obscure relevant complexity, uncertainty, and interdependencies between probability and severity, as when the anticipated severity of an event affects its probability.

Snyder et al. (2020, pp. 11–21) offers a complementary approach to conceptualizing risk that distinguishes between attackers' and defenders' perspectives and sheds light on the relationship

---

[13] We discuss the terms *risk mitigation* and *control* along with the NIST framework in more detail in Appendix E. We recognize that *control* has a specific meaning in the cybersecurity community, as in U.S. Air Force Instruction 17-101 (2020), but we use the word as it used more generally across communities.

between probability and severity by mapping an attack path with Boolean logic statements. Along the path, an attacker must have sufficient access, knowledge, *and* capability to be able to attack *and* must anticipate sufficient impact to choose to attack. Thus, an attacker's decision about whether to attack and, hence, the probability of attacking might be predicated on the expected impact and, hence, anticipated severity. This expression of interdependence between probability and severity points to a mitigation opportunity that we return to later: that acting to reduce impact—or improve resilience—can reduce risk both directly (through severity) and indirectly (through probability).

Figure 2.3. Simplified Risk Matrix

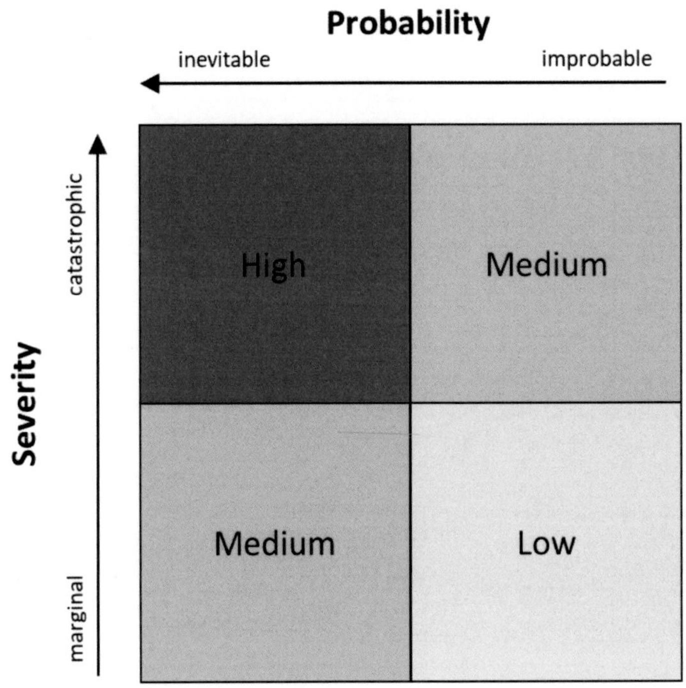

SOURCE: National Academies of Sciences, Engineering, and Medicine (2018, p. 148), based on Greenfield and Camm (2005) and U.S. military guidance.
NOTE: High (red), Medium (orange), and Low (yellow) signify risk levels and possible priorities.

The NIST (2018b) delineation of cybersecurity functions, which we discuss in Appendix E, further highlights not just the interrelatedness of probability and severity but also the potential role of preattack measures to promote resilience. The NIST framework covers five functions: identify, protect, detect, respond, and recover. Identification, protection, and detection could be said to align most closely with preventing an attack or its successful launch; response and recovery speak most directly to reducing the severity or impact of the attack. That said, activities undertaken in the context of identification, protection, and detection can affect impact too. Insomuch as fostering resilience ex ante (before the initiation of an attack) would reduce impact and, thus, lessen the need for response and recovery ex post (after the attack), we can think of resilience as a preemptive contributor to response and recovery.

Chapter 3. The Particular Challenges of Cyber SCRM

In this chapter, we consider the particular challenges of cyber-related risks for supply chains and, hence, SCRM. Past research has tended to explore the distinct properties of cyber-related risks, especially the technological dimensions, more extensively in relation to IT and weapon systems than in relation to supply chains per se (see, e.g., Snyder et al., 2015, and Snyder et al., 2020). As noted in Chapter 2, the literature on cybersecurity tends to focus on the security of information, but the literature on SCRM tends to home in on the surety of supply, with some overlap and the occasional confluence of the two to form a "cyber SCRM" literature. Thus, we begin the chapter by culling from RAND and other research that addresses cybersecurity and SCRM together or separately to identify noteworthy—unique, exceptional, or reinforcing—attributes of cyber-related risks, including the threat environment, that stand out as potentially relevant to SCRM.[14]

Next, we use the lenses of a risk management framework and Boolean attack model (see Appendix E) to consider what makes cyber-related risks different for SCRM. In this regard, we consider complete events, including the initiation of the attack, its outcomes, and its impacts or damage. We recognize that a cyberattack might play out differently from other threats and hazards but that the results could be similar, which might have implications for mitigation. Whether a cyberattack, arson, wildfire, or flood sidelines a facility for a month, the net effect on production at that facility might be the same.

Noteworthy Attributes of Cyber-Related Risks

Here, we catalog various attributes of cyber-related risks, first as they emerge from research on SCRM and, second, as they manifest in work on cybersecurity. The literature offers a panoply of potentially concerning risk attributes that we attempt to synthesize as data points to inform our analysis.

Research on SCRM, referring largely to the surety of supply (e.g., O'Connell et al., 2021), points to a range of physical and nonphysical supply chain risks that can surface unintentionally or intentionally (i.e., with malicious intent). The unintentional sources of risk, which we refer to as *hazards*, could include a range of natural phenomena or disasters, mechanical or technical failures, human errors, technology or policy changes, and market shocks. The maliciously

---

[14] This chapter does not review either literature but rather presents our findings on the noteworthy attributes of cyber-related risks that were evident in the literatures.

intended sources of risk, which we refer to as *threats*, could include cyberattacks, bombings, arson, and product tampering.[15]

Although malicious intent is not unique to cyber concerns, it poses reinforcing challenges by introducing both new channels for *exploitation*, as we discuss later, and a behavioral component (Chapter 4) that can complicate mitigation greatly. As is true for cyberattacks and other forms of threat, defenders must consider how an attacker will adapt its behavior when defenders seek to address the threat (Morral and Jackson, 2009; National Academies of Sciences, Engineering, and Medicine, 2018).

The general SCRM literature also depicts or typifies supply chains as *networks*, even if it does not always deploy this vocabulary (see, e.g., National Academies of Sciences, Engineering, and Medicine, 2018, among many others). Thus, this literature draws attention to the salience of network structure to cyber SCRM and suggests a need to better understand how the attributes of cyber-related risks could play out in a networked environment. We discuss network structure and effects in greater detail in Chapter 5, arguing that a supply chain might present an ideal environment for cyberattacks, which seem able or likely to flourish as a result of interconnections.

Turning to the literature on cybersecurity, Snyder et al. (2020, pp. 6–9) and Snyder et al. (2015, pp. 6–8) highlight several potentially relevant attributes of cyber-related risks to weapon systems and missions. For the most part, their observations raise technological concerns about the threat and its environment, including the following:

- Technologies and capabilities evolve rapidly.
- Cyber-related risks are ubiquitous in time and space, in wartime or peacetime, in deployment or stationed at home, and in design or operation.
- Inherently complex systems breed embedded flaws and vulnerabilities, only some of which can be found and addressed by the defender.
- Cyber-related risks have no firm underlying "laws of nature" or physical presence with which to inform detection or measurement.[16]
- Cybersecurity and systems' functionality are intertwined and potentially at odds, e.g., allowing communication among systems or remote operations could open the door to cyberattacks across systems.
- Interconnected systems can transmit vulnerabilities and limit the separability of decisions on risk mitigation among systems.[17]

---

[15] See e.g., Department of the Army Pamphlet 385-30 (2014). To further illustrate the distinction, accidental or naturally occurring combustion from sparking equipment, bad wiring, a dropped match, or a lightning strike would arise from a hazard, but arson would constitute a threat, even if any or all these events resulted in the same outcome, i.e., a fire of particular scope and intensity with identical damage.

[16] Snyder et al. (2020) draws a comparison with ordinary physical threats.

[17] Snyder et al. (2015) frames this, to paraphrase, as partitioning decisions on risk mitigation being difficult if systems are interconnected or serve as elements of larger systems, but transmission underlies the concern.

Systemic Concerns

Giving a name to concerns about intertwined and interconnected systems, Welburn and Strong (2022) discusses the potential for systemic risk from *common cause* failures, involving commonly held vulnerabilities, and *cascading* failures, involving interdependent networks. In neither case would traditional redundancy—e.g., adding a backup component, duplicating a system, enlisting an alternative supplier, or creating geographic distance—assure resilience. Taking the argument further, mere duplication could make matters worse. In terms of Boolean logic (see Appendix E), adding a component that replicates a preexisting cyber vulnerability could make attacking easier or less costly by increasing the number of potential access points, or the *attack surface*, which could increase the probability of a successful attack. In effect, two systems—or suppliers—with commonly held vulnerabilities can live in the same digital floodplain even if they exist thousands of miles apart.[18] Worse still, interconnected systems can effectively adopt each other's vulnerabilities if an attack on one allows access to the other. It can be as though they have dug irrigation tunnels, so that when one system floods, all the connected systems flood too.

Stealth, Attribution, and Response

Mission failure can come from a loss of command and control, absent any immediately observable component failure.[19] For example, inserting malicious code into a navigation system could eventually lead an aircraft off course without having slowed the production of the aircraft, impeded the delivery of the aircraft, or affected the aircraft's ability to fly on course prior to the code's activation. Similarly, it might be possible to inflict damage on a supply chain without affecting its immediate functionality or attracting observation.

Relatedly, Libicki (2009), Davis et al. (2017), and Romanosky and Boudreaux (2021) address problems of imperfect detectability and attribution. If an attack cannot be detected readily, the damage can unfold over time and space, either through accretion or by delayed release, as in the case of course-altering code. Moreover, if a defender cannot identify its attacker, the defender might be unable to retaliate, credibly threaten retaliation, or guard against future attacks. That said, even if retaliation were possible technically, it might not be desirable in some circumstances. For example, retaliating might entail tipping one's hand about one's defensive tactics, capabilities, or vulnerabilities or might risk further escalation. Over time, cyber observers (e.g., Greenberg, 2020) have argued that cyberattacks have become normalized and, by and large, have received little to no retaliation.

---

[18] Cyber assets can have technologically correlated risks, irrespective of geographic proximity.

[19] This point has come up repeatedly in discussions with our colleagues.

Visibility, Affordability, and Feasibility

Lastly, cyberattacks might offer malicious actors unique, or at least exceptional, advantages over other forms of attack as a matter of visibility, affordability, and feasibility. We us the term *visibility* to capture the extent to which a defender can see a cyber event unfolding, such that visibility might be "low" at the point of initiation—or preceding it, with prior failed attempts—or at points thereafter, even as damage accumulates. If a cyber operation, starting with the attempt, is exceedingly difficult to detect (hence having low visibility from a defender's perspective) and attribute and is low cost or easy, an attacker can try repeatedly to reach a target, because it will not be discovered, punished, or drained of its resources.[20] On that basis, the odds might be stacked in favor of a cyberattacker because it can keep plugging away until it succeeds. Moreover, in some instances, it might have been nearly impossible to reach a target by nondigital means. Imagine a team of James Bond–like operatives attempting to physically infiltrate parts manufacturers across multiple sites, possibly spanning continents, to inflict damage, and compare that with a remote hacker seeking a virtual back door through a common platform. In that case, the costs of an operation—e.g., of obtaining access—might be prohibitive without a cyber option.

Broad Characteristics of Cyberattacks

The foregoing risk attributes are wide-ranging, but the characteristics of a cyber event, consisting of the initiation of the attack and its aftermath, can be binned according to a manageable set of broad characteristics that can be used to compare the event with other nondigital events. In Table 3.1, we have identified four characteristic *types* (onset, duration, visibility, and reach) that describe how events might occur and unfold, each by matter of degree. The first two types, *onset* and *duration*, represent different facets of the pace of a cyberattack. With this typology, we attempted to cover the waterfront of technological attributes both succinctly and in a manner relevant to SCRM. After trying many different arrangements, however, we do not claim that it is the only way to bin the attributes.

Table 3.1. Typology of Characteristics

Characteristic Type	Can or Do Events Occur and Unfold . . .
Onset	Rapidly or gradually and potentially with delay?
Duration	In a fixed interval or indefinitely?
Visibility	With high or low observability?
Reach	Locally, e.g., within a contained system or geography, or across systems and, potentially, globally?

---

[20] Herr (2017, pp. 86–107) points to the ease with which an attacker can strike, but we discuss costs of attacking in greater detail in Chapter 4, raising the possibility that attacking my not be especially low cost.

The characteristics can be bundled to define the contours of an event. For example, an unobservable attack might hit rapidly, persist indefinitely, and reach globally.

In addition, the threat environment is relevant, especially its propensity to change. For example, the threat environment might be dynamic if attackers behave adaptively or strategically (henceforth referred to as *strategic attackers*) or because technological developments occur rapidly.[21] Indeed, rapid change can favor an attacker or a defender. An attacker might be able to exploit rapid technological change to its advantage, but so too might a defender, if, for example, it can update its system protections frequently or continually to block entry to the systems.[22]

Table 3.2 illustrates how the four broad characteristics can cover the range of technological concerns raised at the beginning of this chapter, by mapping the attributes to each characteristic and designating primary relationships with bold double "++" symbols. However, we acknowledge the difficulty of delineating and prying apart some relationships. For example, a condition of low detectability clearly suggests low visibility, but low visibility might, in turn, enable delayed release, persistence, and extended reach, depending on when the attack first comes to light. Thus, for attributes that impinge on visibility, we have designated the secondary relationships with a plain single "+" symbol.

In Table 3.2, ubiquity in time and space (row 1) suggests that an attack can occur anytime (column 2), with potentially long-lasting (column 3) and far-reaching (column 5) effects, e.g., if the incursion occurs in the design or manufacture of a product. By contrast, complexity (row 2), intangibility (row 3), unimpaired functionality (row 4), and practical nondetectability (row 5) all speak directly, almost tautologically, to visibility (column 4) but still have implications for onset (column 2), duration (column 3), and reach (column 5). An unseen force can launch immediately, gradually, or with delay and can spread quietly within and among systems, irrespective of geography.

In this section, we have walked through a litany of proposed unique, exceptional, or reinforcing attributes of cyber-related risks and mapped them to a smaller set of overarching, defining characteristics. But we close with a caution about a risk in our analysis, that of a documented tendency of researchers to seek validation from citations, without aid of direct empirical support.[23] By implication, some commonly held "truths" in the literature might not stand up to empirical scrutiny. For example, constant change is not universal, insomuch as IT might update regularly, but OT can remain in place without substantial alteration for decades.[24]

---

[21] Here, we are contemplating three attributes—malicious intent, the ease with which cyber weapons can be developed and proliferated, and the extent to which technologies and capabilities evolve rapidly—that might all affect the propensity of threats—or the threat environment—to change or take new forms.

[22] See Chapter 4 for a fuller discussion of this point.

[23] This is sometimes known as the "Woozle effect," referring to A. A. Milne's fictional work. For insight into the term's origin, see Kessel, 1995, in a volume dedicated to William Bevan.

[24] The concern pertains to the OT and, in some instances, to the software that supports it. See, e.g., NDIA (2017) and Snyder et al. (2020, p. 6).

For an OT system, longevity could be the main driver of cyber vulnerability. Similarly, ubiquity might be a matter of perspective in that cyberattacks—unlike explosive and chemical attacks, for example—are confined to the digital realm, even if they can affect the physical realm.

Table 3.2. Cyber Threat Attribute Map

Attribute	Characteristic Type			
	Onset	Duration	Visibility	Reach
Cyber-related risks are ubiquitous in time and space	++	++		++
Systems are inherently complex and breed embedded flaws and vulnerabilities	+	+	++	+
Cyber-related risks have no firm underlying laws of nature or physical presence	+	+	++	+
Mission failure can come from loss of command and control, without component failure	+	+	++	+
Cyberattacks might be exceedingly difficult to detect or attribute	+	+	++	+
Cybersecurity and systems' (or supply chains') functionalities are intertwined				++
Interconnected systems can transmit vulnerabilities				++
Cyberattacks can yield cascading failures among interdependent networks				++
Cyberattacks can yield common cause failures among commonly held vulnerabilities				++

SOURCES: Author analysis of attributes that derive from Davis et al. (2017), Herr (2017), Libicki (2009), Romanosky and Boudreaux (2021), Snyder et al. (2020), Snyder et al. (2015), and discussions with colleagues.
NOTE: Bold double "++" symbols indicate a primary relationship, and a plain single "+" symbol indicates a secondary relationship.

Differences Between Cyber and Conventional SCRM Concerns

Cyberattacks stand among many other sources of risk, as depicted in Figure 1.1, but how do they differ? Here, we set aside concerns about intentionality to consider how cyberattacks can unfold as events and inflict damage in relation to an array of conventional hazards, consisting of floods, earthquakes, wildfires, and infectious disease. In so doing, we can isolate the role that cyber-specific technological concerns play in imparting risk and the ways in which it could persist, even if there were no bad actors or malicious codes, just bad accidents and glitches. To draw out the comparison, we examine the hazards in relation to our typology—onset, duration, visibility, and reach (Figure 3.1). This is not intended as a scientific dissection of the hazards, which would require deep subject-matter expertise in each realm, but rather as a means of teasing out differences.

19

On that basis, we found that cyber events can look like the best and, notably, the worst of the rest, largely because of properties of visibility and transmission over time and space. A cyberattack might be confined to the digital realm but can result in damage that occurs immediately, gradually, or with delay; that is possibly unobserved; and that can travel great distances across systems that share platforms or exchange information rapidly.

Figure 3.1 depicts the hazards notionally and simplistically on a continuum for each characteristic. Earthquakes and infectious disease serve as bookends for all except reach. For example, we imagine that an earthquake, traveling along a fault line, sending out seismic waves, and potentially upending an ocean, might outdistance a flood or, possibly, a wildfire, depending, of course, on the specifics.[25] In recent years, wildfires have been covering huge tracts of land, disabling power grids, and affecting distant areas with smoke, but sometimes an earthquake affects only a small area and a handful of structures.

Figure 3.1. What Could Hazards Look Like?

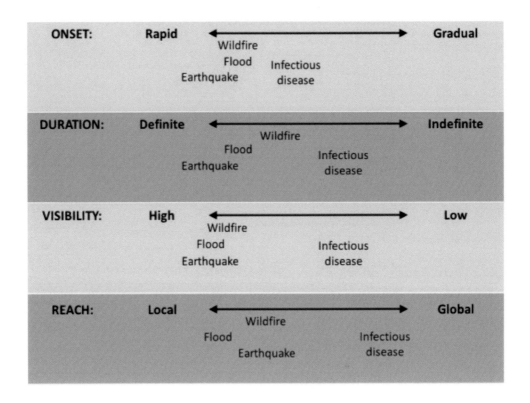

Being at the left of the continuum need not imply a "better" circumstance than being at the right. For example, a rapid onset might be more difficult to counter than a gradual onset, unless

---

[25] Having written much of this report in 2020, we recognize with hindsight that our placement of infectious disease on each continuum was influenced by our experience with the coronavirus disease 2019 (COVID-19) pandemic. For example, we were reading daily reports on the risks of asymptomatic transmission.

the slower pace enables an event to become more destructive either invisibly or without recourse. Similarly, as we noted previously, dynamism can play out to the favor of attackers or defenders.

Even with its ambiguities, the layout of Figure 3.1 suggests that these hazards, except for infectious disease, tend to be highly visible and tethered to time and space, at least in comparison with cyberattacks, which might be said to have nearly free rein. Tying these observations back to our concerns for SCRM, one might argue that the risks associated with these hazards could also be more visible and better tethered, with implications not just for the probability of particular outcomes but for their severity and potential mitigations.

Along similar lines, we noted a distinction in the bounds of the mechanisms and the manifestation of the hazards that bears on the difficulty of comparing their risks against those of cyber threats. To illustrate, the mechanisms of floods, earthquakes, and wildfires are varied but approximately enumerable. For example, in the case of floods, one could point to heavy rain, a broken valve or pipe, or malfunctioning protection systems as among the routes to rising water. A risk analyst could, with help from technical experts, make a list of many if not all the different opportunities for water damage in a given location and assess their probabilities.[26] Cyberattacks, instead, can unfold and wreak havoc in many, many ways, some might say innumerably, even if only digitally. Drilling a layer down from broad classifications shows that the options are immense, perhaps bound largely by imagination, as evident in a large body of taxonomical efforts. (See Ettinger, 2019, p. 111, for a multidimensional effort that covers considerable terrain.) Arguably, the dynamism of the threat environment contributes to the problem of infinitude. An attacker's responsiveness—limited, perhaps, by its creativity and resources—can expand the range of the possible, as can the ease and pace of technological development and proliferation.

Thus, the state of risk assessment in the digital realm might be lagging the state elsewhere, possibly reflecting the depth of uncertainties and a related lack of data, as well as the behavioral considerations that we address in Chapter 4.[27]

Research, reporting, and commentary on the state of risk assessment in the still emerging cyber insurance industry (Box 3.1) point to these and other challenges (see, e.g., Cybersecurity and Infrastructure Security Agency, 2018; Blosfield, 2019; Nissen et al., 2018; Organisation for Economic Co-operation and Development [OECD], 2017; Romanosky et al., 2019; Cyberspace Solarium Commission, 2020; and Wang, 2019).[28] Cyberspace Solarium Commission (2020, pp. 79–80), for example, notes an "inability on the part of the [cyber] insurance industry to

---

[26] Insurers have been dealing with these issues for generations, although apparent increases in episodes of extreme weather conditions could make assessments much more difficult.

[27] For a related discussion of challenges of risk assessment and, especially, data scarcity in addressing terrorist threats, see, e.g., National Academies of Sciences, Engineering, and Medicine (2018).

[28] For more on the analytical challenges, see also Cyberspace Solarium Commission (2020). Nissen et al. (2018, p. 47) suggests creating a data repository at the Department of Homeland Security to help insurers create standardized policies.

comprehensively understand and price risk" that speaks partly to the nascency of the industry but also to the quality of data with which to underwrite policies. In the commission's view (p. 78), "existing datasets are incomplete and provide only a superficial or cursory understanding of evolving trends in cybersecurity and cyberspace."

Box 3.1. Lessons from Markets for Cyber Insurance

The challenges the cyber insurance industry faces might also serve to illustrate the limitations of proposed private-sector endeavors to address cyber-related risk. Cyber insurance has garnered substantial interest in the policy community as a tool for addressing risk pre- and postattack (see, e.g., Cybersecurity and Infrastructure Security Agency, 2018; Cyberspace Solarium Commission, 2020; Nissen et al, 2018; O'Connell et al., 2021; and OECD, 2017), but such insurance may provide limited utility from DAF's perspective, even with efforts to obtain better data and work with the data more effectively. Here, we consider two reasons by describing certain circumstances in which cyber insurance could either undercut DAF's interests or have little bearing on DAF's interests.[a]

First, some enduring data-related problems could encourage overpurchasing of insurance and underinvesting in security, or the opposite. For example, insurers may tend to underprice risk because of inevitable lags or incompleteness in reporting or because they underestimate the risk of infrequent and, especially, costly cyberattacks (see Welburn and Strong, 2022).[b] Alternatively, Bandyopadhyay, Mookerjee, and Rao (2009) describes how contract overpricing can occur when policyholders understand their own risk better than insurers do.[c] Lower prices may result in suppliers overpurchasing cyber insurance, leaving too much to chance, and underinvesting in security, which they might already tend to do (see Chapter 4). Higher prices might, instead, lead to underpurchasing cyber insurance and overinvesting in security. The former might be more concerning for risk reduction, but neither circumstance—too little nor too much investment—is ideal.

Insurance policies can also be priced in relation to a business's apparent vulnerabilities, e.g., through discounts for good hygiene and security investments, or can be written to require that a business take certain actions to reduce its risk, such as regarding security protocols (see, e.g., OECD, 2017), which could help offset a tendency to underinvest. However, the success of a tailored approach to pricing will hinge on the adequacy of the information that is available for judging businesses' vulnerabilities and actions. As discussed elsewhere in this chapter, such information may be lacking and may be especially difficult to obtain in an environment in which the businesses themselves have limited insight into their circumstances.[d]

Second, regardless of pricing, suppliers' and DAF's interests might diverge in ways that leave DAF wanting.[e] An insurance policy that addresses a supplier's financial losses could provide liquidity that relieves constraints on response or recovery but still do little to assure restoration of mission-critical services (e.g., Marotta et al., 2017). Even an operationally oriented policy may leave DAF with disruptions, depending on how quickly restoration occurs and how the supplier prioritizes DAF among its customers (see, e.g., Cybersecurity and Infrastructure Security Agency, 2018). For example, if DAF accounts for only a modest share of a supplier's total business, it might not be sufficiently important from the supplier's perspective to be an immediate beneficiary of service restoration. Although it might be possible to rank DAF higher in contract language, such a provision might be hard to enforce, might run up against complex subcontracting arrangements, and would likely add to the price of the contract. O'Connell et al. (2021) describes ways in which the private sector can reset its priorities through postevent auctions but notes that this and other contractual options to convey flexibility might not be permissible or attainable under the regulations that govern defense contracting.

In the first instance, relating especially to possible underpricing, insurance might undermine suppliers' incentives to adequately invest in protection, which, as we address in Chapter 4, might already be insufficient and might make matters worse. In the second instance, the insurance would not make matters worse but might not help DAF either, except through a possible liquidity effect. After an attack, one can expect a supplier to take whatever actions are in its business interests, but its business interests might not align well with DAF's interests, regardless of its insurance status.

Relatedly, Romanosky et al. (2019) reviews various cyber insurance questionnaires to glean insight into insurance rate schedules and pricing methods, finding significant variance in the amount and quality of information collected by cyber insurers. Although the industry has been working to collect, curate, and apply better data (see, e.g., Blosfield, 2019, on efforts at that time), some of its data-related problems may be hard, if not impossible, to overcome (see, e.g., Welburn and Strong, 2022). So long as the underlying dynamics of a threat are unknown, accurate quantitative assessments of risk may remain out of reach.[29]

Framed in terms of its effects on people, an outbreak of a highly contagious infectious disease (e.g., along the lines of coronavirus disease 2019) comes closest to looking like a cyberattack across the characteristics depicted in Figure 3.1 and with respect to boundedness. Without claiming any medical expertise, it seems to us as if the disease can spread rapidly and unnoticed as presymptomatic or asymptomatic; it can linger in a community or resurface later; it can travel great distances without regard to borders; and it can attack anyone at any time, even with only modest exposure.

Concluding Remarks on Analysis

Returning to the risk matrix, we contend that cyberattacks can yield supply chain disruptions as various conventional hazards and nondigital threats do but, in some instances, can do so with greater effect and open the door to new possibilities, with potentially converging implications for severity, probability, and prioritization (Figure 3.2). Cyberattacks can take myriad forms, occur under the radar and over extended periods, and cover ground, in ways that some conventional hazards and nondigital threats cannot. Cyber capabilities might also put some events that were

---

[29] See Snyder et al. (2020, pp. 7–8) for a fuller discussion of this difficulty.

previously off the matrix, as a matter of feasibility, on the matrix and, thus, increase the range of the possible.

Figure 3.2. Potentially Heightened Risks of Cyberattacks Compared with Conventional Hazards and Nondigital Threats

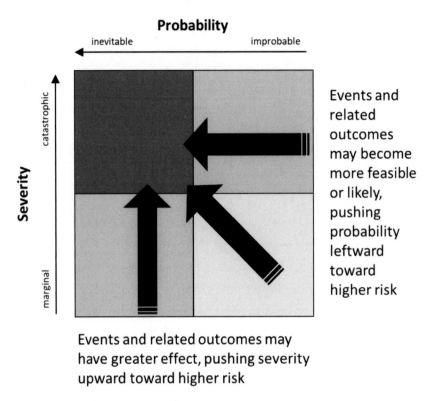

Events and related outcomes may become more feasible or likely, pushing probability leftward toward higher risk

Events and related outcomes may have greater effect, pushing severity upward toward higher risk

SOURCE: Authors' adaptation of National Academies of Sciences, Engineering, and Medicine (2018, p. 148), based on Greenfield and Camm (2005) and U.S. military guidance.
NOTE: Cyberattacks might have more severe consequences; might be more feasible, hence, probable; and might be more severe and more probable, in which case, the forces converge. The colors red, orange, and yellow denote high, medium, and low risks, respectively, and can point to priorities.

In this chapter, we showed that cyberattacks can run the gamut in relation to an array of conventional hazards but, as threats, also present the added twist of intentionality, a dimension that we explore in the next chapter.

Chapter 4. The Implications of Intentionality for Cyber SCRM

In this chapter, we apply principles and findings from the game theory literature (see Appendix B) to explore the role of intentionality in cyberattacks. We do so in light of the influence that the unique, exceptional, and reinforcing attributes of cyber-related risks may have on attackers and defenders and, ultimately, on risk management. However, we recognize that no single game-theoretic model, or set of models, can capture the nuances of the problem, including the intricacies of relationships among actors and their dynamics. Thus, we turned to these models for insight into the implications of intentionality through abstract representations that capture certain aspects of interactions among actors, not for a granular depiction of attacker-defender interactions or their outcomes.

Herr (2017, p. 86) observes that cybersecurity may hold "an air of technical mysticism and opacity," but it is "a very human space." Unlike natural hazards, cyberattacks involve flesh-and-blood attackers with goals, objectives, and strategic capacities, who are acting against and responding to defenders, who have goals, objectives, and capacities of their own. Natural hazards can react—e.g., fire or water can shift positions in response to defensive tactics, and microbes can develop resistance—but they are not, we believe, behaving strategically, as we mean in this analysis.

The problem of intentionality is not specific to cyber-related risk. In fact, it shares features of the challenges that adversarial states, rogues, terrorists, and other criminals' behavior present in nondigital environments. But some of the distinct challenges that cyber-related risks pose, such as those regarding visibility, might either make the problem harder or suggest unexpected trade-offs among—or solutions to—security decisions.[30] Moreover, the specific capacities of cyberattackers, ranging in sophistication from the low-level lone hacker to the well-resourced nation state, might enable them to do things that, while previously imaginable, might have been infeasible. In this chapter, we consider first how this fundamental behavioral component presents defenders, including defense manufacturers, with challenges that traditional, nonadversarial SCRM may not present, then consider the implications of special attributes of cyber-related risks.

Attackers' and Defenders' Decisions in a Game-Theoretic Context

In traditional risk assessment, efforts to mitigate the largest hazards in a system work toward reducing that system's overall risk, without concern for behavioral responses, but threats born of

---

[30] A large literature has formed around this general class of problem that builds on the foundational work of Schelling (1960), Dresher ([1961] 2007), and others who originally focused on nuclear strategy and deterrence (see Appendix C).

strategic interaction will shift as intelligent and strategic attackers find ways around defenders' risk management strategies. Managing cyber-related risks to supply chains consequently pits defenders against attackers in a game of strategic interaction in which defenders seek to minimize the probability of a cyber event and its damage, while attackers seek to maximize them. Table 4.1 provides a few simple examples of how an attacker might respond to a defender's actions (in the form of "if the defender does X, then the attacker does Y") in physical and digital environments. As portrayed in the table, a *defender* might be a government agency seeking to use a policy mechanism to protect public interests, or a business, seeking to protect its interests. Although not depicted explicitly in Table 4.1, a defender might also respond to an attacker, e.g., in a defender-attacker-defender sequence of events (see Appendix B).

Table 4.1. Examples of Strategic Behavior

	Defender	Attacker
Physical threats	• Government imposes restrictions on public access to dynamite	• Bomb maker replaces dynamite with explosive precursor chemicals
	• Manufacturer reduces likely impact of explosion on industrial plant, e.g., by fortifying structure	• Saboteur recalculates benefit-cost trade-off and chooses a different plant
	• Government heightens security at airports	• Terrorist shifts attention to shopping malls
Cyber threats	• Government requires cordoned systems with heightened security for sensitive work	• Hacker seeks system entry through seemingly unrelated external service provider
	• Contractor institutes two-factor authentication to prevent unauthorized access to financial data	• Hacker uses social engineering to bypass enhanced security and gain data access

In the game-theoretic context, the defender's and attacker's moves often come down to making resource allocation decisions in the face of strategic behavior. Specifically, the defender decides how much to invest in security and, if defending multiple targets, where to invest, and the attacker decides how much to spend on an attack and, if choosing among multiple targets, where. In NIST (2018b) parlance, defenders face decisions about allocating investments toward security—primarily *identify*; *protect*; *detect*; and, sometimes, *respond* (see Appendix E)—across multiple possible targets; attackers face decisions about allocating resources toward attacking one or many targets. Each knows that the other might react to the decision in a way that could undermine its objectives. Across threat environments, cyber or otherwise, adversarial relationships can make it harder to guard against risk and may lead to underinvestment in security relative to what would be best for a system, taken as a whole, or even for the business making the decision.

On the difficulty of guarding against risk, protection may reduce risk in relation to conventional hazards, but a strategic attacker may simply seek out the next best opportunity. If a supplier invests in security at one site, a determined and strategic attacker will target another site with less security. Alternatively, some attackers may be opportunistic, compromising vulnerable sites once they discover them. While differing in approach, both types of attacker—determined and strategic versus opportunistic—compromise sites with less security. On that basis, security investments may not be able to guard against all attacks, suggesting some inevitability of *an* attack, but if defenders place their investments strategically, they might still be able to use them to shift attackers' attention away from high-value targets and toward "easier," less-impactful ones.

That said, in the absence of coordination, a defender may find it in its best interest to prioritize certain locations, such as the most obviously vulnerable, at the expense of other locations that hold greater value to national security, or to underinvest from a national security perspective. Underinvestment in security might occur for a variety of reasons—a business might choose to free ride on the investments of other businesses in its supply chain network or to just match the investments in the weakest link in its supply chain network if the supply chain will ultimately be compromised when the weakest, easiest target is compromised (e.g., Bandyopadhyay, Jacob, and Raghunathan, 2010; Nagurney, Nagurney, and Shukla, 2015; Simon and Omar, 2020). Of course, misaligned incentives under a lack of coordination could also lead to overinvestment in cybersecurity. (See also the extended discussion of altering incentives in Appendix C, "Game-Theoretic Models of Cyber Risks to Supply Chains.") These conditions suggest a role for coordination among suppliers, especially given the potential for an attack at one location to negatively affect others, which is a point that we take up in Chapter 5.

As a brief aside, under- and overinvestment can occur outside the game-theoretic context. For example, businesses may underinvest from a societal standpoint simply because their incentives differ from those of national security agencies or, as we discussed in Chapter 3, may underinvest, even from their perspective, because they lack sufficient insight into their risks to make appropriate decisions.

On the problem of incentives, businesses typically strive to meet objectives for profitability, rates of return, etc.; national security agencies, instead, typically work to fill public needs (see, e.g., Nissen et al., 2018, and O'Connell, et al., 2021). In some instances, their objectives might be mutually reinforcing, leading to mutually satisfying investment decisions, but not in all instances. If investments in security measures appear to undermine profitability, businesses might tend to underinvest from DAF's perspective (Fiksel, 2015). Drawing an analogy from a commercial environment, Japanese automakers invested in parts standardization and supply chain mapping after the 2011 earthquake but did not choose to carry additional inventory of parts at factories, because it would render them uncompetitive (Tajitsu, 2016). The automakers might have acted appropriately for their objectives, but a steel mill fire and another earthquake both resulted in production stoppages for Toyota in 2016 (Tajitsu and Yamazaki, 2016). Such a stoppage in defense production might be unacceptable from a national security agency's perspective, but imposing requirements for additional mitigation might drive out suppliers, which would run counter to conventional SCRM concerns about disruption, or might lead to higher costs.

Returning to the game-theoretic context, a defender might underinvest in security, even in relation to its best interests, not just because it lacks insight into the probability or severity of attack, as proposed in Chapter 3, but also because it cannot see how far its supply chain reaches beneath it. Facing an infinite number of possible targets, a defender might not allocate resources to new sites as they come onboard (Bier, Oliveros, and Samuelson, 2007). While the prospect of infinite targets might seem implausible for an industrial supply chain, an equivalent circumstance might arise if a defender does not know how many possible targets it needs to protect because it does not know how far its supply chain reaches. O'Connell et al. (2021, pp. 25–26) discusses limitations on data for mapping the supply chains of specific weapon systems, noting that, "a contractor can only report what it knows, and commercial firms often have little insight more than two tiers down from themselves."

Cyber-Specific Concerns About Information and Visibility

How information travels and gets used may have implications for how—and with what advantages—"games" between adversaries and defenders unfold in the cyber context that differ from those in other contexts. Moreover, cyber-related problems of visibility, as in the detection of an attack, might compound the problems of asymmetric information that are embedded in any adversarial relationship. Despite their best efforts at reconnaissance, both attackers and defenders tend to know more about their own postures—vulnerabilities, resources, and exploits—than they do about their opponent's postures.[31]

Efforts to share information among defenders may lead to gains in efficiency and efficacy, both defensive and productive, but may also present cyber-specific downside risk, and hence trade-offs, depending on the form of the sharing.[32] Within the context of the cybersecurity of industrial supply chains, information-sharing could mean sharing cyber threat intelligence between defenders in the forms of vulnerability information or threat information, sharing or signaling defensive posture and offensive capability between attackers and defenders (see Welburn, Grana, and Schwindt, 2023, and the extended discussion in Appendix B), and sharing

---

[31] In general, information asymmetry would lead an agent, a defender or attacker, to know more about their posture than about another agent's posture. However, in cyberspace, as elsewhere, there are examples of an attacker gathering more information about a defender's systems or circumstances than the defender already possesses about its own systems or circumstances. For example, a defender that cannot see its entire supply chain, as discussed earlier, might not discover vulnerable software in its supply chain before an attacker discovers it. That said, we assume the usual asymmetries apply to agents in this section.

[32] For more on the benefits of sharing, see, e.g., Hausken (2007), Nagurney and Shukla (2017), and Bier, Oliveros, and Samuelson (2007), for more on vulnerabilities, see, e.g., Bandyopadhyay, Jacob, and Raghunathan (2010).

intellectual property (e.g., designs or specifications) between firms. Of particular concern are the possibilities of revealing information that an adversary can exploit to its advantage or of expanding the attacker's attack surface by conveying additional points of entry that increase the probability of successful attacks. While sharing cyber threat intelligence in the form of threat information (e.g., adversary tactics, techniques, and procedures) is unlikely to reveal exploitable information or expand the attack surface, sharing vulnerability information between defenders can entail risk if attackers can gain access to the information and if it provides them with insight, potentially including mere knowledge of a deficit, that they can use advantageously to adjust their tactics or plan future attacks. Sharing information about specifications between firms may enhance efficiency or improve quality in the supply chain—it might also reduce the risk of disruption, if, for example, it enables more businesses to step in as alternative suppliers. However, sharing that information can also undermine security by adding new vulnerabilities that would convey additional points of entry and, thus, expand the potential attack surface.

Undetectability or limited detectability presents another challenge in the interplay between defenders and adversaries, particularly regarding the potential for retaliation and deterrence. Most obviously, it is harder to deter an adversary if the adversary knows it faces little or no prospect of punishment because it is hard to detect (see Chapter 3). However, the possibility of more-complex strategic behavior on the part of the attacker could add another layer of concern. For example, less-aggressive attackers may benefit from hiding in the shadows of more-aggressive attackers, which can drive or increase aggression among the less aggressively inclined.[33] In an environment in which one cyber adversary is known to be particularly aggressive, defenders are more likely to just assume that that adversary is the source of most attacks, allowing lesser-known adversaries to benefit from false attribution. The behavior is roughly analogous to that of motorists who maintain speeds just below those of the fastest motorist with some confidence they will not get pulled over.[34]

The literature suggests that some challenges of information and visibility can be mitigated partly with strategies that either play off asymmetries, possibly by adding to them, or compensate for them. For example, a defender might be able to leverage the asymmetry to deter an attack by holding information on the extent of its investment in security close or creating confusion about its circumstances. Defensive strategies based on secrecy and deception with respect to information and posture applied over time may sometimes present a more cost-effective security strategy than candor (see, e.g., Zhuang, Bier, and Alagoz, 2010). In addition, when poor visibility impedes detection, a defender may be able to benefit from layered defense strategies and from partial and randomized (i.e., unpredictable) detection strategies (see, e.g., Jackson and

---

[33] For a formal example, see, e.g., Baliga, de Mesquita, and Wolitzky (2020).

[34] However, even when detection and retaliation are possible, the literature suggests that deterrence strategies that either publicly name and shame or retaliate in kind may be effective, but only some of the time against some adversaries. For more on retaliation with imperfect detection and attribution, see, e.g., Edwards et al. (2017) and Welburn, Grana, and Schwindt (2023).

LaTourrette, 2015, and Haphuriwat, Bier, and Willis, 2011, respectively). A layered defense strategy might, for example, entail adding a second or third identification to a log-in process and a protective firewall. In the case of partial and randomized strategies, a defender can gain from adding some uncertainty to the attacker's equation, roughly analogous to its ability to leverage asymmetric information.

Are Cyberattacks Cheaper or Easier Than Others?

Cyberattackers might also benefit from both lower expected costs of retaliation, given weak detectability, and lower costs of attack in comparison with attackers employing noncyber means, depending on the circumstances. A single bomber, for example, might have only one or few opportunities to attack. In comparison, a cyberattacker can repeat its attack many times, without a high probability of detection or retaliation. Cyberattacks may therefore be seen as more feasible than other forms of attack.

Still, the near certainty of "a" cyberattack does not make all attacks possible or even desirable for an attacker. On feasibility, an attacker must be able to plan and act fast enough to avoid detection *and* remain technologically current. If a defender can refresh its technology in less time than it takes for an attacker to gain access, knowledge, and capability and to have an impact (i.e., the Boolean attack model from Snyder et al., 2020), the defender might be able to deter the attack. In effect, a strategy of technology refresh would make attacking costlier by forcing an attacker to operate more rapidly, although, at some point it might not be possible to operate rapidly enough.[35] A technology refresh strategy, might, however, be more expensive for some defenders than others, depending on the range of technologies they employ. For example, if a manufacturer replaces machinery only every ten or 20 years, switching out a technology—or even attempting a modest upgrade—could entail a substantial financial loss that could, if large enough, drive some businesses out of the market.[36] Regarding desirability, cyberattacks might be less costly for attackers—as a matter of coordination and, possibly, financing—than other options in some circumstances but, even then, are not necessarily cheap or free.

Consistent with this Boolean perspective, one might look to cost-raising measures to reduce the probability of attack, but the prospect of doing so raises three related questions: "How much does it cost to stage a cyberattack?" "What are the drivers of those costs?" and, by extension, "What can be done to make it costlier for the attacker?"

---

[35] Arguably, a security-oriented strategy could be described similarly.

[36] For more on OT life cycles and implications, see NDIA (2017). An economist could frame the problem in terms of capital depreciation rates, implied capital turnover, and long-run profitability. If a manufacturer's business model is built around an assumed depreciation rate on machinery that implies replacement once every ten or more years, a defensively oriented investment strategy that requires a faster refresh and shorter service life would effectively raise the depreciation rate and reduce the manufacturer's long-run profitability. If the refresh were presented as a compliance requirement, the costs could drive some firms out of the industry.

In Box 4.1, "Uncertainty of the Costs of Attacking," we present two different approaches to estimating the costs, the results of which illustrate the challenges of answering these questions. We find that the results differ substantially, depending on the estimation method and circumstances, and that anyone seeking to pursue a cost-based strategy would need better information about an attacker's "business model" to assess its merit (Greenfield and Paoli, 2013, p. 866).[37] Huang, Siegel, and Madnick (2018, p. 3), for example, developed a "cybercriminal value chain model" that includes costs of attacking, but the differences in valuation that emerge from their and other approaches to cost estimation suggest that much remains to be learned about how attackers' costs accumulate and in what amounts. Any effort to make attacking costlier for an attacker would also entail costs for the defender, with attendant business implications. Whether the effort would matter—such that the benefits to the defender outweigh the costs—would depend, in part, on the attacker's modus operandi.

Concluding Remarks on Analysis

Our exploration of game theory and cyber SCRM suggests four takeaways:

- Defenders can increase security investment and possibly discourage attacks at one location, but in so doing, they may encourage them elsewhere. This suggests the impossibility of preventing all attacks, but defenders might be able to redirect attackers to targets of lesser importance and reduce risk overall.
- Defenders may choose to protect just some locations or to free ride on others' security investments, leaving some targets vulnerable, which may present special challenges in a networked environment, such as a supply chain. Although defenders, acting independently, may misdirect or underspend on security from a societal perspective, coordination could lead to societally preferable investment decisions. However, defenders may still underinvest in security, even in relation to their interests, if they cannot see how far their supply chains reach.
- Defenders must contend with informational asymmetries, compounded by cyber-specific visibility challenges, but may be able to gain from sharing information among themselves, if they can avoid increasing their exposure to risk. However, under some circumstances, defenders may also benefit from secrecy and deception, e.g., by not revealing their security posture or creating confusion about it.
- Defenders can also try to make attacks less attractive by making them costlier or harder (e.g., by threatening retaliation or refreshing technology) but may encounter substantial obstacles. For example, defenders cannot credibly signal high retaliation costs, unless they can overcome difficulties of detection and attribution, and might not know whether they are raising costs enough to matter.

---

[37] Greenfield and Paoli (2013) define the *business model* as a structure that "depicts the typical logistics or *modus operandi* of a criminal activity . . . ." It can be used to gather evidence and inform the analysis of the harms of the activity and, eventually, the costs and benefits of potential responses to it.

Box 4.1. Uncertainty About the Costs of Attacking

Two back-of-the-envelope approaches to estimating the costs of attacking produce results that differ by six orders of magnitude. The results differ not just by method but also by circumstances and point to the difficulty of implementing a cost-based strategy without better information.

First, one might add up the costs of obtaining everything an attacker would need for conducting a cyberattack or *hack*—e.g., acquiring infrastructure (Deloitte, 2018), developing or acquiring exploits or zero-days (Ablon, Libicki, and Abler, 2014), testing methods, planning the attack, and covering the hacker's time. The costs could range from hundreds to millions of dollars.[a] At the low end, an attack might require just a laptop, off-the-shelf tools, no testing or planning, and a few minutes of a hacker's time. At the high end, it might require considerably more effort, including a fully appointed front company and a lengthy development process. For insight into development costs, we could treat malware development as simply a form of software development, which would fall in about the same cost range, depending on the size and complexity (Huijgens et al., 2017). We could then walk through each of the factors (see Wagner and Ruhe, 2018) that drive software development costs and try to estimate an attacker's cost from these factors or look at rental rates for hacking products. For example, it is possible to rent Mirai for $7,500 to get 1 terabit per second of use (see Bing, 2016) or spend $3,000 to $4,000 to use 50,000 bots (see Mathews, 2016). Along similar lines, Huang, Siegel, and Madnick (2018) takes a value chain modeling approach and estimates investment and operating costs of about $13,000 to $14,000 to run a monthlong ransomware attack.

Alternatively, one could use employment data on U.S. cyber forces and estimate what it would cost to employ the same personnel nefariously at comparable wage rates. As reported in 2016 and 2018, respectively, Cyber Mission Force reached initial operating capability with about 5,000 people spread over 133 teams (see U.S. Cyber Command, 2016) and was expected to be fully operational with about 6,200 people spread over the 133 teams (see U.S. Cyber Command, 2018). One offensive mission team, of which Cyber Command has 27, employs 64 people to conduct operations and 39 people in support roles, such as target discovery, analysis, language, and malware analysis (Caton, 2015; Pomerleau, 2017). Employing 6,200 people at $100,000 each, which could be a low estimate for highly skilled, technical personnel, would cost more than half a billion dollars annually, without accounting for any supporting equipment or infrastructure. This suggests that cyberattacks are not "cheap." By comparison, an offensive mission team, amounting to about 100 members, would cost about $10 million per year for labor alone, which is considerably less but still in the millions of dollars.

The higher implied costs of the employee-based estimate might derive from an underlying difference in cost structure that may or may not make sense in the cyberattack context. The teams are stood up and trained and equipped over several years, which implies more sunk than variable costs. That structure could make sense for hackers if, for example, attacks have intrinsic training value. Consequently, attacks might be conducted—as a matter of training—even if they would not seem to make financial sense on their own.

Without a better understanding of the attacker's business model and the main drivers of the costs of cyberattacks, it might be unrealistic to imagine that raising the costs, to the point of affecting an attacker's decisionmaking, is an attainable goal. Our preliminary efforts to decompose and estimate costs suggest that those seeking to implement a cost-based deterrence strategy would need more information to do so effectively.

[a] The categories might overlap. For example, it might not be easy to separate the costs of the hacker's time and those of tool or malware development. Planning, testing, and software development can all overlap too.

These takeaways suggest a range of challenges for cyber SCRM and, in some instances, opportunities to reduce risk, relating to investment incentives and information-sharing. For example, the value of encouraging attackers to shift targets to less worrisome sites stems from the impossibility of preventing all attacks. Moreover, the tendency to underinvest from a national security perspective and individually may find a partial remedy in coordination. Yet, even without concerns about adversarial behavior, differences in businesses' and national security agencies' objectives, as well as a lack of insight into underlying risks, could present obstacles to achieving appropriate levels of investment. On that account, a program like the DoD's

Cybersecurity Maturity Model Certification (CMMC) program,[38] which was nascent at the time of our research, could provide a means of achieving something that looks more like a coordinated outcome and that might stand to reconcile some differences in interests—but not without cost to the businesses that participate. At the margin, any program that requires businesses to increase their spending could discourage some from entering the market or lead some to exit the market, possibly by pursuing commercial opportunities. Insomuch as a program results in fewer defense suppliers, it could, albeit unintentionally, undermine more-traditional efforts to reduce the risks of disruption. In addition, we found that sharing some forms of information, honestly or deceptively, can reduce risk and improve quality or efficiency but can also present trade-offs if it increases the potential for exposing sensitive information that attackers can use to their advantage.

Finally, a supply chain is not just a series of multiple vulnerable targets; rather, supply chains are networks of their own networks. Each location on the network represents a different supplier with different incentives and potentially different information. The result is a game that is made up of multiple defenders in which problems of information-sharing and coordination abound and can multiply. We discuss some of the implications of maintaining security in a networked environment, such as a supply chain, in Chapter 5.

---

[38] The CMMC program, according to publicly available documentation at the time of our research, was intended as a unifying standard for implementing cybersecurity across the defense industrial base that provides assurance to DoD that a contractor can protect sensitive information and, through cascading provisions, can account for the flow of information down to subcontractors within a multitier supply chain. The program, as initiated in version 1.0, required a firm to verify—through certification—the implementation of the processes and practices associated with the achievement of a cybersecurity maturity level that is intended to be commensurate with its role (Office of the Under Secretary of Defense for Acquisition & Sustainment, undated).

Chapter 5. Interactions Between Cyber-Related Risks and Supply Chains

In this chapter, we pick up on a theme that we introduced in Chapter 3 and consider ways in which network attacks on supply chains are, effectively, network attacks on networks. To start, we review findings from the literature on network analysis, pertaining largely to production networks and disasters, that point to the relevance of interdependencies among suppliers for our research. The literature works with a method known as *input-output* (IO) analysis because it can be used to trace the relationships between and among businesses, sectors, and economies through their use of inputs and production of outputs. Next, we draw out the implications of interdependencies among suppliers, analyzing the effects of different types of cyberattacks on a series of highly stylized supply chains. We present the basic features of the stylized supply chains and then explore instances of disruptive and exploitative attacks in each. The results, which we describe in a narrative form in this chapter and mathematically in Appendix D, shed light on how the structure of networks, consisting of relationships among suppliers and flows of products and, potentially, information, alters cyber-related risks to supply chains.

We recognize here, as in Chapter 4, that our renderings represent substantial departures from reality and make no claim to capture all the workings or nuances of actual defense industrial supply chains. In those supply chains, relationships do not look like trees, consisting of discrete actors. To call out just a few variations, relationships can be circular (see Appendix C), businesses can act as their own suppliers, and some products may travel across international borders as imports or exports. (For depictions with these and other more-complex features, see also National Academies of Sciences, Engineering, and Medicine, 2018.) Our intent, drawing on a rich literature in network analysis (see also Appendix D), is to strip the underlying concept of a *supply chain* to its bare essentials to home in on specific points of concern and extract high-level insights about structure and vulnerabilities.

Why Do Interdependencies Matter?

The literature on production networks cautions us that interdependencies among businesses, sectors, or other locations of industrial activity matter because they can serve as conduits for and amplify the effects of shocks through potentially widespread ripple effects. A shock, as through a natural or maliciously induced disaster, at one location in a network can affect not just the immediately adjacent locations but can also reverberate beyond them, possibly throughout the entire network. Thus, the literature sheds light on how a cyberattack on one supplier in a defense industrial supply chain can travel within the supplier's own business networks—both IT and physical—and up and down a supply chain. Insomuch as an ordinary, physical network might serve as a conduit for transmission, an attack on a network within a network might be said to further amplify the effects. The literature also suggests the relevance of a supply chain's structure and the shock's form in that different structures or types of shocks may respond or behave differently.

To illustrate, Santos and Haimes (2004) uses an IO model to study the extent to which the economic effects of a terrorist attack that reduces airline demand can propagate to other sectors. Santos and Haimes (2004) finds that the disaster risk of one sector, in this case air transportation, can create significant risks for others through connections among the sectors. In the terrorist attack scenario, the effects travel as far as oil and gas extraction, implying the potential for repercussions not just at the point of attack but well beyond. Santos, Haimes, and Lian (2007) extends that approach to consider the economic impact of a cyberattack on supervisory control and data acquisition systems, also finding widespread effects.

Acemoglu et al. (2012) provides a general mathematical framework, without reference to a specific hazard or threat, for understanding how isolated shocks can pose large aggregate risks in networked environments. In their application, the authors found that idiosyncratic shocks to the microeconomy can propagate upstream and downstream and, thus, affect the macroeconomy. In related work, Acemoglu, Ozdaglar, and Tahbaz-Salehi (2015) finds that the impact of shocks to networks depends on the structure of the underlying network, with some structures being more resilient to small or large shocks than others.

Others have looked more specifically at the role of firm-level production networks and the propagation of disruptions to individual firms and their customers. For example, Barrot and Sauvagnat (2016) examines the propagation of shocks through supply chain relationships after natural disasters. The authors found that, following disruptions, businesses pass a significant share of their losses on to their customers, which, through further propagation through supply chains, can lead to significant economic loss. Boehm, Flaaen, and Pandalai-Nayar (2019) focuses on the aftereffects of the 2011 Great East Japan Earthquake and, in addition to finding confirming evidence of propagation, presents evidence of the difficulty businesses faced finding new suppliers with which to recontract after the earthquake.[39]

Welburn and Strong (2022) extends this literature, focusing on the potential systemic risks of cyberattacks. The authors used an IO framework similar to those of Santos, Haimes, and Lian (2007) and Acemoglu et al. (2012) to model the cascading effect as a hypothetical cyberattack leads to losses that spread upstream and downstream through industrial supply chains. Taking a more empirically oriented approach, Crosignani, Macchiavelli, and Silva (2021) uses data on firm linkages to identify the propagation of disruptions within the supply chain networks following the 2017 NotPetya cyberattack. Both approaches found that most of the supply chain damage following a cyberattack comes from propagation to downstream customers. Furthermore, Crosignani, Macchiavelli, and Silva (2021) found that, even if businesses were

---

[39] See also MacKenzie, Barker and Santos (2014), Carvalho et al. (2020), and others.

unable to find substitute suppliers in the near term, the NotPetya cyberattack led to adjustments in customer and supplier relationships in the long term.

In addition to demonstrating the potential for amplification of a cyberattack in a production network, the literature provides evidence on the challenges of engaging with substitute suppliers—at least in the near term—to enable recovery (see also Chapter 4), the potential for long-lived changes in customer and supplier relationships, and the disproportionate effects of shocks on downstream customers that should be of interest to DAF. In the next section, we introduce stylized supply chains with which we explore the transmission and effects of cyber-related risks of disruptions and exploitation.

Basic Features of the Stylized Supply Chains

A given defense industrial supply chain is defined by a collection of circular *nodes*,[40] representing different firms that are the defense industrial manufacturers that we call *suppliers*, and the linkages between them, representing the flows of products and, potentially, information. The network is directional; that is, firms upstream in the network supply inputs to other firms downstream in the network that eventually use the inputs to produce a finished item (e.g., a weapon system or spare part) for delivery to DAF or another customer. Suppliers that are directly connected to the customer represent *prime contractors*, and suppliers that are indirectly connected represent *subcontractors*. We can also describe the network in terms of tiers, in which suppliers directly connected to the customer make up the first tier, suppliers directly connected to the first tier make up the second tier, and so on.

Applying the vocabulary of attackers and defenders from Chapter 4, each supplier in the network represents a different *defender* with a probability of being compromised by an *attacker* and with a given outcome and impact if it is attacked. To focus our discussion on the role of network structure, we start by making the simplifying assumption that all suppliers have equal importance in the chain, such that an attack on any supplier would have the same impact (we discuss the implications of relaxing that assumption in subsequent analysis). Furthermore, we differentiate between the risks of cyberattacks for supply chain disruption and exploitation. In this setting, a disruption to the supply chain can result in a loss of product availability or quality or an increase in cost, while exploitation, consisting of infiltration and exfiltration, could result in a loss of product integrity or information, which could also bear on availability or quality and cost.[41]

We, therefore, describe the stylized supply chain accordingly, as a set of suppliers, directional linkages, and probabilities. Later, we use this construct to estimate the risks of

---

[40] Throughout this report, we adopt the language of network analysis in which *node* (or, equivalently, *vertex* in graph theory) refers to a single entity connected to other entities in a network through *arcs* (or, equivalently, *edges* in graph theory).

[41] See the related discussion on the relationship between integrity and quality in Chapter 1.

disruption and exploitation and to draw general insights into the nature of cyber-related risks to supply chains and potential mitigations.

Risks of Disruption and Exploitation

We will start by estimating the probability of disruption for the three stylized supply chains, shown in Figure 5.1 as Cases A, B, and C. In the leftmost supply chain, Case A, disruption can come from an attack on either the single second-tier supplier (denoted as 2) or the prime supplier (denoted as 1). The same is true for the rightmost example, Case C, but an attack on a third-tier supplier (denoted as 3) adds an additional opportunity for disruption. Case B has two suppliers, but instead of lengthening the chain, as Case C does, it adds a substitute (denoted as 2′), which implies that a disruption to this second tier would require a successful attack on *both* suppliers (2 and 2′).[42] Thus, Case B requires node 2 or node 2′ for a complete supply chain, while Case C requires node 2 and node 3. Equivalently, the supply chain in Case B is disrupted if node 2 and node 2′ are disrupted, and the supply chain in Case C is disrupted if node 2 or node 3 is disrupted.

Figure 5.1. Basic Stylized Supply Chain Cases

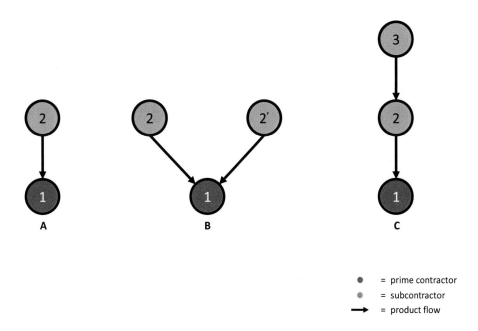

NOTE: This figure displays three basic stylized supply chain cases: In Case A, supplier 2 supplies inputs to supplier 1; in Case B, suppliers 2 and 2′ supply substitute inputs to supplier 1; and in Case C, supplier 3 supplies inputs to supplier 2, which, in turn, supplies inputs to supplier 1.

---

[42] As a practical matter, this could mean that the consumer purchases all goods from one supplier or the other but can switch suppliers instantaneously or that it purchases some from each but could immediately expand its purchases from one or the other to meet its needs.

Here, we summarize the results of our analysis in Appendix D, where we formally calculate the risk of disruption and exploitation for each case in Figure 5.1.[43] In the event of attempted disruption, the results are broadly consistent with the traditional perspective that adding redundancy—in the form of alternative suppliers—can support supply chain resiliency. However, the possibilities of exploitation, particularly exfiltration, and of shared vulnerabilities underscore concerns raised in Chapter 3 about the limitations and implications of redundancy so defined.

First, all things being equal, the risk of disruption for Case B is lower than that for Case A, which is lower than that for Case C. Two features drive this result: Adding tiers to lengthen a chain increases the risk of supply chain disruption, while adding substitute suppliers within a tier reduces the risk of disruption.

Second, we considered risks of exploitation, but, for simplicity and as a point of stark contrast, focused on exfiltration, such as the theft of a design. If all suppliers have the same information, such as the design for a final product, an attacker that seeks to extract that information from the supply chain would only require access to one supplier.[44] Then, for exfiltration, we found that, relative to the two-tier supply chain (A), adding a redundant supplier (B) or another tier (C) increases the risk of exfiltration by expanding the attack surface and increasing the risk of exfiltration by the same amount. Simply put, each additional supplier gives the attacker more opportunity in equal measure.

Considering what would happen if the value of information—and hence the potential impact of attack—differs among tiers introduces some nuance. As less or less-valuable information is shared further upstream (e.g., at third-tier suppliers instead of second-tier suppliers), its exfiltration may become less harmful to the ultimate customer, such as DAF. This assumes that information decays (or with some mathematical equivalence, that the probability of exfiltration declines) with each upstream connection, which alters the overall probability, and thereby risk, of exfiltration.[45] Then, as we show in Appendix D, the marginal risk of adding a tier decreases with each added tier, which could imply that a supplier at an upper tier needs less protection than a supplier at a lower tier.[46] This result points to the possible value of adding redundant suppliers at tiers further upstream.

This benefit can be illustrated through a simple example. Consider the stylized three-tiered supply chains shown in Figure 5.2. Supply chain Case A serves as a baseline for comparison as a

---

[43] As we explain in Appendix D, with the assumption of equal impact (and, hence, severity), calculations of probability and risk are equivalent in our simple examples.

[44] For purposes of infiltration, an attacker might still need to reach all suppliers.

[45] See Appendix D for details on this and other assumptions.

[46] If, however, in an alternative scenario, information grows in value upstream rather than decays, the opposite might be true. For example, an upstream supplier may have essential information about the design of a subcomponent that cannot be obtained elsewhere. Information growth, in contrast to decay, would point to the value of adding redundant nodes at tiers closer downstream.

supply chain without any redundant suppliers. Both supply chains B and C introduce redundancy with a single alternative supplier, but supply chain B adds the supplier to the second tier, while Case C adds it to the third tier.

Figure 5.2. Supplier Redundancy in Stylized Supply Chains

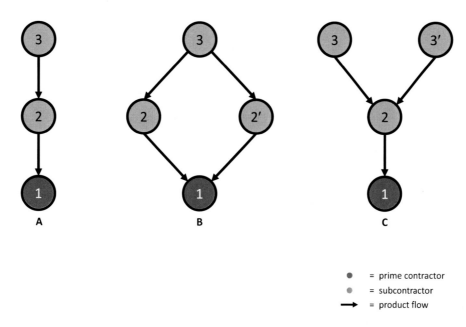

NOTE: This figure displays three ways of introducing redundancy in a three-tier supply chain. Supply chain A presents a baseline supply chain with just three suppliers; supply chain B introduces a redundant supplier in the second tier; and supply chain C introduces a redundant supplier in the third tier.

Using the same approach as previously, we calculated the risk of disruption and exfiltration for the three examples in Figure 5.2 (see Appendix D). We found that, while both B and C have a lower probability of disruption than A, the risks of disruption for B and C are equal. While a similar statement of the equivalency of supply chain Cases B and C could be made for exfiltration, the inclusion of information decay alters the picture. In the case of information decay, the risk of exfiltration is higher for B than it is for C. Therefore, in the presence of information decay, or declines in the value of information by tier, the added risk of exfiltration from redundant suppliers can be partially offset by adding suppliers further upstream in the supply chain rather than downstream.

Importantly, a lack of independence among suppliers significantly alters these results. Consider the simple examples of the two-tier supply chains shown in Figure 5.3. Our calculations in Appendix D show that, in comparison with supply chain Case A, supply chain B, with redundant suppliers in the second tier, has a lower risk of disruption but a higher risk of exfiltration. However, if the second-tier suppliers (2 and 2′) are not independent, as shown in Case C, and share a common disruption-allowing vulnerability (i.e., a common cyber floodplain) that could be exploited to compromise both with the same attack, this example has both a higher risk of disruption and a higher risk of exfiltration.

Figure 5.3. Supplier Independence and Interdependence

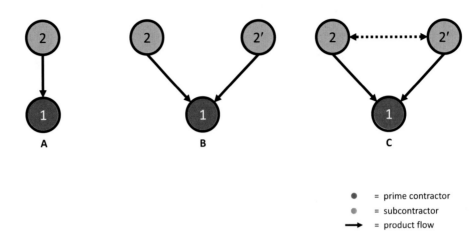

NOTE: This figure displays three stylized two-tier supply chains. In supply chain A, supplier 2 supplies inputs to supplier 1; in supply chain B, suppliers 2 and 2′ provide substitute inputs to supplier 1; and in supply chain C, suppliers 2 and 2′ are also dependent on common software, shown with a dashed line.

Concluding Remarks on Analysis

Our review of the literature on production networks lends weight to concerns about network attacks on networks and points to some additional challenges, beyond those of amplification. For example, businesses may have difficulty finding alternative suppliers in the near term, which could impede recovery, and downstream customers may bear the brunt of any shocks. With the supply chain analysis described in this chapter and the mathematical representations in Appendix D, we can draw four general insights, each of which we support formally in Appendix D. All things equal, we found the following:

- Adding a tier to the supply chain may add to risks of both disruption and exploitation, suggesting that risk mounts with deeper supply chains, especially, as findings from game theory suggest, if the length of the chain is uncertain.
- Adding a redundant supplier to a single tier may decrease the risk of disruption while increasing the risk of exploitation, suggesting the possibility of trade-offs among risk-reduction objectives and options.
- Adding a redundant supplier upstream rather than downstream can add less risk of exploitation if information decays with successive tiers, suggesting that, although such risk mounts with supply chain depth, it may do so at a decreasing rate.
- Adding a redundant supplier reduces disruption risk only if the probability of a successful attack on the redundant supplier is independent of the probability for others; if it is not independent, this redundancy can add to both disruption and exploitation risk. Thus,

common technological vulnerabilities among suppliers, as through shared systems or like platforms, can undermine any benefits of adding suppliers.

These findings highlight dual concerns about supply chain configuration, including uncertainties regarding its contours, and the nature of redundancy. In the context of conventional SCRM, redundancy often refers to capacity, as might be reflected in the number of suppliers at any given tier in a supply chain, spread across locations; in cybersecurity, it typically refers to a technology. However, in the digital realm, a technology, to be truly redundant, must be separate and, preferably, distinct.[47] This is not to say that conventional SCRM has no interest in technological independence; working with separate and distinct technologies that can fill a common need can reduce the risks of disruption outside the digital realm by reducing the risks of single points of failure in production or distribution. Nevertheless, redundancy can mean something different in each context and, by extension, the implications of adding or lacking redundancy in each context can have very different risk implications. Adding redundant suppliers, in the duplicative sense, can increase risks of exploitation, while lacking redundant— separate and distinct—technology can increase risks of disruption and exploitation.

A deficit of knowledge about the composition of a supply chain and commonly held technological vulnerabilities within supply chains might represent the worst of all possible worlds but could also represent the world we live in. O'Connell et al. (2021, pp. 25–26), for example, discusses the defense community's lack of visibility beneath the second tiers of what the authors describe as increasingly complex defense supply chains and the limitations of the data that are available to improve visibility. The best real-world examples of the pitfalls of common vulnerabilities may come from back-to-back ransomware attacks in 2017. Both WannaCry and NotPetya plagued a wide range of organizations, from health care, to education, to manufacturing and distribution, after quickly spreading from infected machine to infected machine across the world (see, e.g., Collins, 2017). Maersk, the global shipping giant, reported losses of as much as $300 million during NotPetya (Thomson, 2017). The scale and reach of the damage from the attacks was due largely to the fact that both exploited EternalBlue, a widely held vulnerability across Windows operating systems. Both episodes serve as sound reminders of the lack of independence across suppliers, where commonly used software exposes seemingly separate suppliers to shared risks.

---

[47] See also the discussion of Hausken's (2008) findings on redundancy in Appendix C.

Chapter 6. Conclusions

This report encapsulates our efforts to characterize cyber-related risks to supply chains for defense industrial hardware and to identify directions for risk assessment and mitigation and for research. At the outset of this report, we posed two related research questions: "How do cyber-related risks differ from or compound other concerns about SCRM?" for defense industrial supply chains and "What do, or could, these differences mean for risk assessment, risk mitigation, and research?"

On the first question, cyber threats are just one source of supply chain risk, but, as our research shows, they could be worse than and different from other sources of risk. Cyber events can present the worst of all the characteristics of conventional hazards, including earthquakes, floods, and wildfires, by occurring suddenly and spreading rapidly across digital and supply chain networks that can span continents, potentially unobserved and with instantaneous or slowly mounting consequences. Under these conditions, a cunning adversary with intent to harm may be able to obtain outcomes that were not feasible previously, possibly at low cost and with little concern for punishment. At the same time, the risks of cyberattacks might be unhindered or even elevated by some conventional approaches to addressing cyber and SCRM concerns separately.

On the second question, we found that cyber SCRM requires a comprehensive bearing that encompasses concerns not just about the security of information but also about supply chain functionality, including the availability, quality, and cost of deliveries. Moreover, if an attack is inevitable, cyber SCRM must give due weight to response, recovery, and resilience. It must also recognize differences in DAF and industry interests that could affect whether and how industry contributes to risk reduction and account for potential trade-offs among risk-reduction objectives. Borrowing from Paoli and Greenfield (2015), DAF might gain from orienting its objectives for risk assessment and mitigation to "start from the end," by homing in on consequences for mission attainment to set priorities. We also suggest areas of research that can support cyber SCRM, by delving into the details of some long-standing issues and exploring new ones.

In the remainder of this chapter, we present the analytical insights that yielded these findings and elaborate on directions for cyber SCRM and research.

Analytical Insights

In this section, we present a set of insights that draw from our analyses in each chapter, speak to our research questions, and set the stage for a fuller discussion of directions for risk assessment, risk mitigation, and research. In Chapter 2, we introduced our lines of effort and, in the remaining chapters, explored the results. In Chapters 3, 4, and 5, we used different methods to characterize—and differentiate—risk, but the methods also revealed needs in risk assessment, risk mitigation, and research. In Chapter 3, we compared the challenges of cyber-related risks with those of other threats and hazards and found differences in pace, visibility, and reach that bore on risk assessment and mitigation. In Chapter 4, we considered interactions between attackers and defenders, related vulnerabilities, and potential responses. In Chapter 5, we analyzed the interaction between cyber-related risks and supply chain structure to shed light on the risks from disruption and exploitation and the trade-offs among risk-reduction objectives.

The first insight, as shown in Table 6.1, pertains directly to our first research question, and the subsequent insights pertain largely to our second research question. There is some overlap among the insights and in their contributions to each question. Table 6.1 also connects each insight into supporting evidence chapter by chapter.

Table 6.1. Crosswalk Between Analytical Insights and Sources of Evidence by Chapter

Analytical Insights in Relation to Research Questions	Noteworthy Attributes (Chapter 3)	Findings from Game Theory (Chapter 4)	Findings from Network Analysis (Chapter 5)
How do cyber-related risks differ from or compound other concerns about SCRM?			
Damage to supply chains from cyberattacks could be worse than and different from damage from other threats or hazards	++	++	++
What do, or could, those differences mean for related risk assessment, risk mitigation, and research?			
Preventative measures are not enough	++	++	++
Cyber SCRM requires more than an amalgam of "cyber" and "SCRM"	+	+	++
Private-sector efforts to manage risk may not meet national security needs	++	++	++
Research can go deeper and further to support cyber SCRM	+	++	++

NOTE: Bold double "++" symbols indicate a primary relationship, and a plain single "+" indicates a lesser role. Chapters 4 and 5 draw from materials presented in Appendixes B and C, respectively.

Damage to Supply Chains from Cyberattacks Could Be Worse Than and Different from Damages from Other Threats or Hazards

Our characterization of the cyber landscape led us to conclude that the damage to supply chains from a cyberattack could be worse than and different from the damage from other sources of risk. When we assessed noteworthy attributes of cyber-related risks in terms of the potential rate of onset, duration, visibility, and reach of a cyber event, we found that an event could manifest as a worst case in relation to a set of conventional hazards and could pose even greater challenges than nondigital threats. Cyber events can take myriad forms, occurring suddenly and spreading rapidly across digital and supply chain networks that can span continents, potentially unobserved and with instantaneous or slowly mounting consequences. Thus, they can strike under the radar and over extended periods and can cover ground in ways that conventional hazards and nondigital threats typically cannot.

The potential for cyberattacks that are less costly—or easier to undertake—than their nondigital alternatives and for repeated unpunished attempts might also increase the range of the possible. In our James Bond–like example, a team of operatives would need to physically infiltrate parts manufacturers across multiple sites, possibly across continents, to succeed, but a single hacker might only need to access one virtual back door remotely. The cyberattack would not be costless (see Box 4.1) and might require more than a one-person effort, but it might be more amenable to execution and less costly than a nondigital version. Adding to the concern that the fallout from cyberattacks could be worse than and different from that of other, nondigital sources of risk, our exploration of game-theoretic models reminds us that an strategic attacker can be expected to strike the hardest at the worst time and seek to do as much damage as possible. Furthermore, our stylized supply chains shed light on the potential for attacks to travel and spread damage throughout supply chains, especially considering the possibilities of poor visibility—into both the attack and the composition of the supply chain—and commonly held vulnerabilities among suppliers.

Preventative Measures Are Not Enough

Analysis throughout the report makes a strong case against relying too heavily on prevention, especially if efforts to prevent attacks come at the expense of efforts to facilitate response and recovery or build resilience. Creating impenetrable defenses is both infeasible and inherently difficult. Irrespective of feasibility, attempting impenetrability is generally inadvisable because the attempt would entail risks and costs of its own. We do not claim this caution against overreliance on preventative measures as a novel insight but rather as a well-founded one. It is based not just on our analysis but is also supported by the literature (e.g., Cyberspace Solarium Commission, 2020, Defense Science Board, 2017, and Bartock et al., 2016), by fundamental tenets of risk management (e.g., Greenfield and Camm, 2005), and by recent cyber events.[48] In our analysis, the insight rests largely on our characterization of cyber-related risks, including the threat environment, our exploration of defenders' and attackers' behavior and its implications for cyber SCRM, and our evaluation of the network properties of supply chains.

---

[48] See, for example, Bartock et al. (2016, p. vi), which states, "There has been widespread recognition that some of these cybersecurity (cyber) events cannot be stopped and [that] solely focusing on preventing cyber events from occurring is a flawed approach." The NIST document advocates for improvements to prevention capabilities and detection and response capabilities, but the emphasis is on recovery.

On the difficulty of prevention, we found earlier that cyberattacks can take nearly innumerable forms; can spread widely without detection; and can yield damage immediately, incrementally, or with delay. We also identified persistent challenges of risk assessment that relate to data deficiencies and, perhaps more fundamentally, to the depths of uncertainties and the seeming unboundedness of the problem.

Turning to behavior, we discussed how increasing security at one location can lead an adversary to attack a different, less-protected location. As we note later, this might create an opening to deflect attention from one site to another in a way that would reduce risk overall but also means that an attack would still occur. Defenders may also leave some sites unprotected or may free ride; in a networked environment, such as a supply chain, doing so can leave others more vulnerable. At the same time, a highly motivated adversary can be expected to find an open door, strike painfully, and benefit disproportionally from cyber-specific visibility problems.

By analyzing supply chains as networks, we showed that deeper supply chains can be riskier than shallower ones, especially if their contours are uncertain, as might be said of defense industrial supply chains, and that hidden dependencies, which may be commonplace in a supply chain, can obscure vulnerabilities and thwart mitigation.

Moreover, although our analysis warns against overemphasizing prevention because it is infeasible, conventional thinking on risk management also suggests that it is also inadvisable (Greenfield and Camm, 2005). Trying to drive risk to zero by driving the probability of a successful attack to zero would involve risks of its own, in terms of both the interplay among risks—efforts to eliminate one risk would or could trigger another—and resource costs. Relatedly, we draw attention to a seemingly technical point that we raised in Chapter 2, regarding the interrelatedness of probability and severity.[49] In that chapter, we noted that efforts to reduce the potential impact of an event, such as by taking steps to facilitate response and recovery or to build resilience, could reduce risk both directly (through severity) and indirectly (through probability). Thus, underutilizing measures directed toward improving response, recovery, or resilience may mean passing up an opportunity to reduce risk from two mutually reinforcing directions simultaneously.

The infeasibility and inadvisability of guarding against all risk also suggest a need for prioritization, which, as we will argue, may entail some reorientation toward concerns for consequences—outcomes and impacts—for both information security and supply chain functionality and apart from attacks per se. That is, if DAF must accept some risk of an attack, it should be prepared to deal with the consequences, especially those of greatest criticality. The NotPetya attack, which effectively halted Maersk-operated ports, vessels, and container ships, can provide some insight into the dimensions of concerns for supply chain functionality from a national security perspective. While NotPetya disrupted the transportation and delivery of commercial goods, it is easy to imagine an analogous scenario, affecting the timely delivery of

---

[49] See also Appendix E.

45

parts or weapon systems supporting an ongoing U.S. military operation or the rapid deployment of U.S. troops.[50]

Cyber SCRM Requires More Than an Amalgam of "Cyber" and "SCRM"

Our game-theoretic and network analyses point to shortcomings of approaching cyber SCRM as an amalgam of *cyber* and *SCRM*, including the potential for outright conflict among risk-reduction objectives. On that potential, we found that adding suppliers to allay concerns about disruption can increase the risk of exploitation, especially if the suppliers share vulnerabilities. We also noted how adding cybersecurity requirements for suppliers could result in less market participation, which might increase the odds of disruption and higher costs. Absent any trade-offs, a fusion of cyber- and SCRM-based measures could be inadequate, insomuch as conventional SCRM underestimates the potency of cyberattacks relative to other sources of risk.

Here, we also note a subtle difference in how the cybersecurity and SCRM communities tend to think about information and its role in risk. Cybersecurity might be said to prioritize restricting access to information to prevent leakage and, to a lesser extent, sharing information to aid identification, protection, and detection. By comparison, conventional SCRM might prioritize sharing information to promote collaboration and interoperability, albeit not without concern for preventing leakage.

We also explored differences in the meaning of "redundancy" across cyber and SCRM contexts; specifically, in conventional SCRM, *redundancy* might be associated first, although not only, with additional suppliers; in cybersecurity, it pertains mostly to technology. In neither setting would mere duplication necessarily mitigate risks. In conventional SCRM, one might seek geographical separation, with suppliers spread across distant locales; in cybersecurity, one would seek technological separation. However, technological separation might hold value in both contexts—cybersecurity and conventional SCRM—without involving trade-offs among objectives for risk reduction because some commonly held vulnerabilities can contribute to risks of exploitation and disruption. Moreover, technological separation can serve not just cyber but also conventional SCRM purposes, e.g., by reducing the noncyber risks of single points of failure in production or distribution.

Private-Sector Efforts to Manage Risk May Not Meet National Security Needs

All told, our analysis suggests that private-sector efforts to manage cyber-related risk to supply chains may not be able to meet DAF's needs for information security or supply chain functionality. Strategic interactions between defenders and attackers could lead to underinvestment in security, especially without coordination among defenders, and several compounding factors, involving risk assessment, incentives, and supply chain visibility, could

---

[50] For a discussion of threats to DoD, the defense industrial base, and the U.S. military's ability to deploy forces and project power and influence abroad, see Gonzales et al. (2020). See also Carter (2012).

make matters worse. For example, underdeveloped markets for cyber insurance, owing partly to data inadequacies, coupled with underpricing of insurance, could reinforce concerns about underinvestment in security (see Box 3.1). Moreover, even if the markets were better functioning and if pricing were as it should be, suppliers' and DAF's interests might diverge unhelpfully from DAF's perspective in the wake of an attack. For example, DAF—or a downstream contractor that sells to DAF—might not rank first among a supplier's priorities for order fulfillment after recovery and might not benefit expeditiously from renewed production. Finally, a supplier that cannot see into the depths of its own supply chain or is unaware of dependencies within it cannot be expected to mitigate risk to its own satisfaction, let alone to that of DAF or other customers. Our analysis does not rule out a private-sector role—it is essential—but, rather, suggests that any role must consider potential obstacles and differences in DAF and industry interests.

Research Can Go Deeper and Further to Support Cyber SCRM

Throughout this report, we have used existing research and stylized applications of well-vetted methods to better understand cyber-related risk and to consider implications for directions in risk assessment and mitigation, but we have also uncovered needs to go substantially deeper and further. By *deeper*, we mean delving into the details of issues that have already received some analytical attention, possibly with new or different analytical methods, and, by *further*, we mean exploring issues that we have only just uncovered or confirmed with our analysis.[51] For example, from game theory and network analysis, we learned how defenders can reduce risk by redirecting attackers to targets of lesser importance, how supply chain interdependencies can incentivize underinvestment in security, how coordination among defenders could offset some such tendencies, and how differences in risks of disruption and exploitation can imply trade-offs among risk-reduction objectives. While drawing out those findings, however, we also identified substantial knowledge deficits. Some of these deficits reflect the inherent limitations of the methods of inquiry, including their granularity and tactical relevance, but others reflect the novelty of the questions at hand. Similarly, we found opportunities for advancements in research and methods in other areas, such as those concerning private-sector engagement and risk assessment. Thus, from our work with the current literature and from our analysis in this report, we have found that research, including methods development, can go deeper and further to support cyber SCRM for DAF and its supply chain vendors and, we posit, will be necessary to support the shift that we suggest in the next section, under the "Directions in Cyber SCRM" heading, to a more-comprehensive approach to cyber SCRM.

The first of our analytical insights has led us to conclude that DAF will need to come to terms with the potential fallout of a cyberattack that may not have a nondigital equivalent, and the rest suggest that cyber SCRM might not yet be up to the task of confronting the attack. We

---

[51] We offer specific suggests on methods in Appendix B.

47

have identified a need to fully recognize and balance cyber and SCRM concerns over the full life cycle of a cyber event. Here, we are not suggesting that DAF—or industry—should cast aside defensive measures or disregard opportunities to improve them. However, having found that prevention *alone* is not enough, we are suggesting close consideration of response, recovery, and resilience. In the section that follows, we consider what this could mean for cyber SCRM, as well as research, now and in the future.

Directions for Cyber SCRM and Research

In this section, we focus on our second research question by drawing together implications for directions in risk assessment, risk mitigation, and research. If neither DAF nor industry can gain perfect insight into risk and/or stop every attack, what can be done to address cyber-related risks more effectively, for DAF, and how can research support these efforts? We suggest starting from a mission-oriented perspective by establishing priorities for cyber SCRM in relation to consequences for mission attainment, whether DAF is considering the potential for attacks through or on supply chains, events that it can guard against or must deal with through response and recovery, or some combination of these possibilities. That, in turn, would mean taking on response, recovery, and resilience fully and directly while accounting for the distinct challenges that cyber-related risks present to SCRM for defense industrial products, differences in interests among stakeholders, and potential trade-offs among risk-reduction objectives relating to information security and supply chain functionality. Then, we propose areas of research to support the approach.

Directions in Cyber SCRM

We suggest approaching cyber SCRM comprehensively, reaching from the beginning of a cyber event to its end, with close consideration of the consequences of cyberattacks for the functionality of defense industrial supply chains and mission attainment and of related opportunities to build resilience and restore mission-critical functionality.

In that context, we suggest the following:

- framing the potential consequences of cyberattacks in terms of the availability, quality, and cost of defense industrial products that serve mission-critical roles, not just or primarily in terms of information security
- establishing priorities among those cyber and SCRM consequences based on what they could mean for mission attainment
- setting out terms for cyber SCRM strategies, with due attention to response, recovery, and resilience that account for concerns about

  - information security and supply chain functionality
  - differences in DAF and private-sector interests that could affect whether and how industry contributes to risk reduction

– trade-offs among risk-reduction objectives, relating, for example, to risks of disruption and the availability of alternative suppliers, on the one hand, and the vulnerability of information, on the other.

Our intention is not to abandon concerns for identification, protection, and detection or for information security but to take on cyber SCRM comprehensively and holistically from the perspective of a cyber event's consequences. From that perspective, one might imagine two overlapping and potentially interrelated sets of concerns—pertaining to information security and to the functionality of the supply chain itself—each bearing on mission attainment. Damage in one domain could imply damage in the other, and damage in either domain, separately or jointly, could lead to mission failure. Taking the concerns together in full view of their linkages to each other and to mission attainment, DAF could establish its priorities for addressing the risks of a cyberattack in relation to its mission. In the next section, we outline an approach to setting those priorities.

Commercial mechanisms that could assist in response and recovery—or contribute to building resilience—include insurance and contract language, but not without substantial caveats. Insurance might relieve constraints on liquidity to facilitate response and recovery and provide incentives to improve hygiene or invest in security through discounts or underwriting requirements but might itself encourage underinvestment if underpriced and cannot assure DAF's satisfaction, given differences in DAF and business interests. Contract language might be adjusted for help on recovery, e.g., by adding provisions on DAF's standings among its supplier's customers after an attack, but placing DAF at the front of the line would likely increase the price of the contract and might run up against complex subcontracting arrangements that could complicate enforcement.

Measures to soften the blow of an attack hold the potential to reduce risk directly, by reducing the severity of consequences, and indirectly, by reducing their attractiveness and, hence, the probability of occurrence. In the Boolean attack model, an attacker who expects little reward for their efforts might not bother attacking in the first place. One approach to reducing impact might be to add industrial suppliers that do not have commonly held cyber—or other technological—vulnerabilities. DoD, through the Office of the Under Secretary of Defense for Acquisition and Sustainment and its predecessors, has a long-standing—if mixed—record of considering and shaping policies to promote market entry as means of maintaining or improving the health of the defense industrial base (see, e.g., Brady and Greenfield, 2010, and Lorell et al., 2002), and defense manufacturers commonly maintain relationships with multiple upstream suppliers for surety of access to parts and other production inputs. Our analysis, however, suggests the importance of considering correlations among vulnerabilities in those pursuits.

Although we have warned against overemphasizing prevention, we do not intend to foreclose the possibility of constructive, preemptively oriented interventions, many of which could also involve or depend on private-sector engagement. For example, defenders, including businesses, might be able to make attacks harder or costlier for attackers by obscuring, manipulating, or divulging false information; by leveraging technological change to outpace attackers and deny them access; and by instituting good cyber hygiene and security protocols as a rudimentary line of defense. Even if a defender can only redirect an attacker's attention to a less worrisome target, the defender might still be able to reduce risk overall. Such redirection could constitute a form of prevention in that it would stop an attack on a target of particular concern, but an attack on a different target would still occur. Moreover, although coordination among defenders to address underinvestment or free riding might not occur naturally—i.e., without external encouragement—it could play a part in risk reduction. However, whether any of these efforts would matter and matter enough to merit their expense would depend, in part, on how attackers undertake their operations.

Finally, efforts to develop better methods of assessing risk—as could contribute to analyses of trade-offs among risk-reduction objectives or to fostering a more robust insurance market—or to build stronger defenses against risk should not impede progress on addressing response and recovery. For example, in choosing how to share information or whether to encourage market entry, it would be helpful be able to assess the net effects of potential decreases and increases in the risks of disruption and exploitation. However, without reasonably accurate risk assessments, it is not possible to directly weigh the two risks and potential trade-offs among risk-reduction objectives. Breaking a potential logjam in risk assessment or prevention would mean leveraging the five-step risk management process as a tool for continuous improvement (Figure 6.1), not lingering at Step 2 (assess threats and hazards) or, when venturing beyond, aiming for zero risk.

Current guidance on risk management, as embodied in the five-step process and related documents (see, e.g., AFPAM 90-803, 2017) already encourages this approach, but it could be revised to include a more explicit statement of intent, regarding the continuous nature of the process, the limits of risk reduction, and the implications of accepting residual risk for response, recovery, and resilience, and it might use examples from this, the cyber SCRM, context.

As discussed in Greenfield and Camm (2005, p. 49),

> [r]isk control, which occurs in steps three, four, and five as part of risk mitigation, would involve developing a strategy for eliminating, reducing, or coping with the possibility of a hazard. By implication, the goal of risk mitigation is not necessarily risk elimination. In some instances, it may be preferable to accept some amount of "residual risk" and develop a response and recovery plan.

Thus, mitigation in this context includes responding, recovering, and building resilience, but formal guidance on risk management, such as that found in AFPAM 90-803 (2017), could be clearer about the linkage and draw out the connection between this framework, the NIST functions, and cyber SCRM.

Figure 6.1. Leveraging the Five-Step Risk Management Process

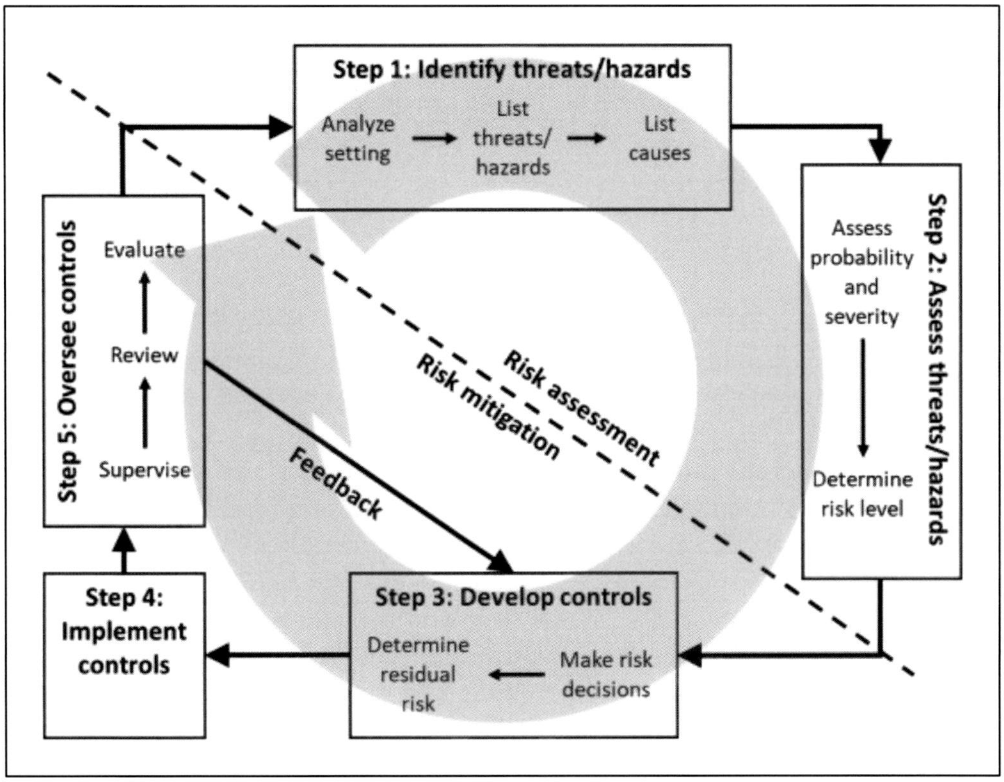

SOURCE: Authors' adaptation of Greenfield and Camm (2005), based on U.S. military guidance.

Directions in Research on Cyber SCRM

If research can go deeper and further to support cyber SCRM, where should it go and how? Here, we identify four areas that pertain to the DAF's interests; in Appendix B, we delve into each, both with further elaboration and by proposing specific research questions and approaches to answering them. The first area pertains to how DAF approaches risk assessment; the second, to establishing DAF's needs and priorities for response, recovery, and resilience; the third, to private-sector efforts to manage risk; and the fourth, to developing a comprehensive cyber SCRM strategy:

- **Approaching risk assessment with realistic expectations and with greater emphasis on supply chain functionality**. If DAF cannot know everything about its cyber-related risks, what should it try to know, and how should it deal with the rest, i.e., true uncertainty? While DAF cannot map the whole ocean, it should be able to identify and parameterize the most salient features of the ocean; along these lines, we recommend pursuing research to probe the breadth and depth of supply chains and to uncover hidden technological vulnerabilities. Such research would focus on establishing the contours of DAF's defense industrial supply chains as they stand today and as they could stand in the future, given the composition of the defense industrial base and possibilities of business entering or exiting the market.

51

- **Establishing needs and priorities for supply chain response, recovery, and resilience**. While establishing needs and priorities is a core DAF function, an evidence-based understanding of needs and priorities for response, recovery, and resilience to cyberattacks would better support policymaking and implementation. We suggest a three-part approach, involving a process of needs elicitation, evaluation, and sorting, that would consider response, recovery, and resilience in the context of a supply chain's role in mission attainment.

- **Examining the utility and limits of private-sector risk reduction**. Differences between DAF and private-sector interests, as well as practical concerns about the data, methods, and visibility needed to support risk assessment, may limit the ability of would-be private-sector solutions to address DAF's concerns. For example, as we uncovered in this report, industrial suppliers' incentives could lead them to underinvest in security relative to DAF or their own objectives. With such concerns in mind, we recommend pursuing research that examines the potential for using incentive mechanisms to improve the alignment of DAF and private-sector interests; for cyber insurance to better contribute to response, recovery, and resilience; for strategic use of information; and for increasing the costs or difficulty of attacks. Among many promising options for exploring these issues, DoD's CMMC program, which had only just commenced at the time of our research, might provide a fruitful case study on incentives, behavior, and outcomes.

- **Crafting a comprehensive strategy for cyber SCRM**. Taking a more-comprehensive approach to developing a strategy for cyber SCRM would require better information about the form and extent of the trade-offs among risk-reduction objectives, but that, in turn, would require improvements in risk assessment that may be unattainable. Given the potential for lasting uncertainty about risk parameters in the cyber domain, a comprehensive strategy for addressing cyber-related risk must, then, also be resilient. Research in this area could test proposed strategies for their robustness under varying conditions.

Some of the research in these areas can serve more than one purpose, for example, by contributing to a better understanding of the risks of cyberattacks and, at the same time, to balancing risk-reduction objectives. In some instances, extensions of the methods that we explored in this report—e.g., game theory and network analysis—might help fill the need and, in others, researchers might turn to entirely different approaches for granularity that would lend itself to tactical solutions. Accordingly, some of the research, such as that involving behavioral modeling, could yield high-level insights but not such tactical granularity or concrete, actionable solutions.

We do not mean to imply that these are the best or only areas to consider for research or intend to imply that they have not yet received any attention but, rather, that our research pointed squarely to them, as needing additional attention.[52] Without more work in each area, it could be difficult to promote readiness in the face of a seemingly unavoidable cyberattack. In

---

[52] Much as we noted previously that the policy community has not stood still in the time since we completed the research for this report, the research community has also continued with its efforts. Thus, we recognize that more work has—almost certainly—been done in each area and is ongoing.

Appendix B, we provide a full discussion of each proposed research area, including potential research questions and approaches to answering them.

Concluding Remarks

In this report, we have discussed how the risks that cyberattacks pose to defense industrial supply chains differ from those of conventional hazards and threats and considered directions for risk assessment, risk mitigation, and research. Although some of what we suggest can begin immediately (e.g., reframing concerns about cyberattacks to better emphasize consequences and priorities in relation to supply chain functionality and mission attainment), change may take time because our recommendations require filling knowledge gaps. While some research might begin to bear fruit in a matter of months, some could take years. Among the former, it might not take long to gather data on the state of play in cyber insurance; among the latter, it could take several years to understand the effects of the CMMC program and, even then, only with a better understanding of baseline conditions. Although setting out a long-term endeavor to confront an immediate threat has obvious shortcoming, it is, nevertheless, our intention that this report provide a foundation for better assuring that DAF gets what it needs, when it needs it, at an acceptable cost.

Appendix A. Definitions

The use of terms of art, such as *Cyber-SCRM*, *cyber SCRM*, and *C-SCRM*, and risk-related related vocabulary, such as *hazards* and *threats*, differs widely across sources, including formal policy statements and military guidance. In this research, we use these terms in particular ways, which we set out in this appendix.

What Do We Mean by *Risk*?

Risk, as we define it in this report, is a combination of the probability of an event, brought on by a threat or hazard, and its severity in relation to potential outcomes and their impacts or damage.[53] For our purposes, a cyber *event* consists of the initiation of the *attack*, including breaches, and its outcomes and impacts or damage. *Probability* in this context consists of both the probability of the attempt of a cyberattack, which might depend partly on the relative attractiveness and expected impact of the attack, and the probability of the success of that attempt. Within cybersecurity fields, this is commonly referred to as the "capability, intent, and access" of the threat actor (*attacker*). Severity, in turn, concerns the impact or damage associated with the set of possible outcomes of the attack. In the context of cyber-related risks to supply chains, the outcomes and damage could involve *disruption* to the supply chain (such that products arrive late, in insufficient quantity, of dubious quality, or at higher-than-expected cost) or *exploitation* through infiltration or exfiltration. In the case of infiltration, an adversary might insert a malicious code that results in a malfunction or improper use; in the case of exfiltration, it might enable espionage by syphoning off sensitive proprietary or national security information.[54]

What Do We Mean by *Cyberattack*?

In many DoD contexts, *attack* carries a precise definition: "Actions taken in cyberspace that create noticeable denial effects (i.e., degradation, disruption, or destruction) in cyberspace or manipulation that leads to denial that appears in a physical domain, and is considered a form of fires" (Joint Publication 3-12, 2018, p. GL-4). A cyberattack would thus differ, for example, from an act of cyber exfiltration that extracts information and leads to a loss of confidentiality. However, across audiences and methods, including those in the disciplines that we used to construct our analysis, the term cyberattack is often used differently or more broadly. NIST, a

---

[53] See Aven et al., 2018, and AFPAM 90-803, 2017, for similar approaches.

[54] We use the term *sensitive information* in this report in lieu of specific terms, such as *controlled unclassified information*, that formally define or categorize different types of sensitive information, because we are speaking more generally and because of differences in use among sources over time.

widely referenced source of cybersecurity standards and definitions, offers the following definition for *cyberattack*: "An attack, via cyberspace, targeting an enterprise's use of cyberspace for the purpose of disrupting, disabling, destroying, or maliciously controlling a computing environment/infrastructure; or destroying the integrity of the data or stealing controlled information" (NIST Computer Security Resource Center, undated).[55]

In turn, NIST Computer Security Resource Center (undated) defines *attack* variously, for example, as "Any kind of malicious activity that attempts to collect, disrupt, deny, degrade, or destroy information system resources or the information itself" and as "The realization of some specific threat that impacts the confidentiality, integrity, accountability, or availability of a computational resource," and in terms of unauthorized entities' deceitful practices.

In this report, we adopt a broad perspective on attacks and attacking in our use of the term *cyberattack* and further specify types of actions or intrusions by their impact on the supply chain, highlighting both disruption and exploitation. We also work with the long-used terms in game theory of *attackers* and *defenders*, referring to agents in adversarial relationships. By implication, we are not using the terms *cyberattack* or *attack* to impart any legal weight or operational authority.

How Do Hazards and Threats Differ?

We define *hazards* as unintentional sources of risk and *threats* as intentional sources, specifically, malicious acts. *Hazards* would include naturally occurring phenomena, such as floods, earthquakes, wildfires, and mechanical failures and human errors; *threats* would include bombings, arson, and product tampering. This approach enables us to compare the risks of cyberattacks with those of other hazards and threats and draw out the implications of intentionality. For other treatments of this vocabulary, see, e.g., AFPAM 90-803 (2017) and Department of the Army Pamphlet 385-30 (2014).

What Is Cyber SCRM?

In the absence of perfect terminology for our purpose, we use the phrase *cyber SCRM* (no hyphen) broadly, to refer to the cybersecurity of supply chains, taken to include attacks on supply chains in which the target of the attack is the supply chain itself, not just the security of cyber supply chains or the information contained within them. Although our emphasis in this report is on concerns for the security of defense industrial supply chains, we take a more expansive view of cyber-related risks to supply chains than typical *Cyber-SCRM* (*C-SCRM* and *CSCRM*) delineations, which often focus largely or entirely on the risks to supply chains that produce IT and related products and services.

---

[55] See the NIST glossary entries for *cyberattack* and *attack* (NIST Computer Security Resource Center, undated).

What Is Resilience?

In this report, we use the definition offered in Presidential Policy Directive 21 (2013): "The ability to prepare for and adapt to changing conditions and withstand and recover rapidly from disruptions. Resilience includes the ability to withstand and recover from deliberate attacks, accidents, or naturally occurring threats or incidents." On that basis, we consider resilience, an *ability*, as related to response and recovery (see Appendix E) but different from response and recovery.[56] A more resilient supply chain might withstand and recover from a cyberattack more rapidly than a less resilient supply chain, but the more resilient supply chain might still require some attention or repair after an attack. In that way, we think of building resilience as an ex ante activity that can reduce impact and diminish the need for response and recovery, both of which occur ex post.

---

[56] Response and recovery are two of five cybersecurity functions in the NIST Cybersecurity Framework (identify, protect, detect, respond, and recover; see NIST, 2018a). For definitions of these terms and an explanation of how they relate to other vocabulary in this report, see Appendix E.

Appendix B. Opportunities for Research on Cyber SCRM

In this appendix, we explore specific opportunities for research in the four areas set out in Chapter 6 (Table B.1):

1. how DAF approaches risk assessment, in terms of both expectations and emphasis
2. establishing DAF's needs and priorities for responding to, recovering from, and increasing the resilience of supply chains to cyberattacks
3. the utility and limits of private-sector risk reduction from DAF's perspective
4. crafting a comprehensive strategy for cyber SCRM that is, itself, resilient to rapidly changing threats.

As noted in Chapter 6, we do not mean to imply that these are the best or only areas to consider, and we do not intend to imply that they have not yet received any attention. Rather, our research pointed squarely to them as needing additional attention.[57]

Table B.1. Research Areas, Illustrative Research Questions, and Possible Approaches

Research Area	Illustrative Questions	Possible Approaches
Approaching risk assessment with realistic expectations and with greater emphasis on supply chain functionality	• How do the parts of the supply chain relate to the whole?	• Case and/or industry studies • IO analysis, computable general equilibrium models, and agent-based models
	• Where are the vulnerabilities, e.g., shared suppliers, systems, and software?	• Case and/or industry studies • IO analysis, computable general equilibrium models, and agent-based models • Data science methods of technology assessments
Establishing needs and priorities for supply chain response, recovery, and resilience	• What needs constitute the highest-ranking priorities?	• Mental models, gaming and tabletop exercises (TTXs), and logic modeling • Empirical validation • Criteria formulation and systematic ranking procedures, based on objective measures
Examining the utility and limits of private-sector risk reduction	• What steps can be taken to better align industry's interests and DAF's needs?	

---

[57] Just as the policy community has not stood still in the time since we completed the research for this report, the research community has also continued its efforts. Thus, we recognize that more work has—almost certainly—been done in each area and is ongoing.

Research Area	Illustrative Questions	Possible Approaches
	– How does supply chain configuration affect investment?	• Game-theoretic, network, and hybrid models of disruption and exploitation
	– Can policy or contract design elicit coordination?	• Mechanism design models
	– What can we learn from the CMMC program?	• Case study
	– What are the implications of differences in IT and OT?	• Capital replacement models
	• What can cyber insurance do for response, recovery, and resilience?	• Industry benchmarking and tracking • Behavioral modeling • Data and methods review
	• What part can information play in risk reduction?	• Game-theoretic, network, and hybrid models of modes and types of sharing
	• What can businesses—or policy—do to raise the costs or difficulty of attacking?	• Cost estimation • Tactical models of defensive strategy • Capital replacement models • Training review
Crafting comprehensive cyber SCRM strategies	• Where and how big are the trade-offs among risk-reduction objectives?	• See entries for risk assessment and needs, and priorities
	• How well does the approach hold up under uncertainty and balance objectives?	• Robust decisionmaking, gaming and TTXs, and pilot programs

Approaching Risk Assessment with Realistic Expectations and with Greater Emphasis on Supply Chain Functionality

If DAF cannot know everything about its cyber-related risks, what should it try to know, and how should it deal with the rest, i.e., true uncertainty? While DAF cannot map the whole ocean, it should be able to identify and parameterize the most salient features of the ocean. Along these lines, we recommend pursuing research both to probe the breadth and depth of supply chains and to uncover hidden technological vulnerabilities.[58] Such research would focus on establishing the

---

[58] This research could complement work undertaken pursuant to EO 14017 (2021), which calls for sectoral supply chain assessments, including the following:

> The Secretary of Defense . . . shall submit a report on supply chains for the defense industrial base that updates the report provided pursuant to Executive Order 13806 of July 21, 2017 (Assessing and

contours of DAF's defense industrial supply chains as they stand today and as they could stand in the future, given the composition of the defense industrial base and possibilities of businesses entering or exiting the market. This research would improve our understanding of these supply chains and the health of the underlying defense industrial base as two distinct but interrelated concepts. The participants in a supply chain deliver a defense industrial product, and the defense industrial base represents the universe of potential market participants.

Looking at specific supply chains that produce mission-critical defense industrial products could help identify choke points or weaknesses from a disruptive or exploitative perspective. For example, ongoing RAND research has found that a wealth of distributors for some products can create the appearance of many suppliers when there are actually very few. Viewed through the lens of network analysis, the risk profiles of the two configurations—with and without many suppliers—would differ greatly, and using that lens might improve the ability to evaluate the differences and consider mitigations (Figure B.1). To illustrate the problem in a nonmilitary context, someone seeking to buy a standard household HVAC filter might find many online sellers, but these sellers could be purchasing from just a few manufacturers and, possibly, using similar distribution channels.

In this area, we suggest looking more closely at classes of models (e.g., IO, computable general equilibrium, and agent-based) that can be used to evaluate the supply chain linkages of the industrial base, simulate the impacts of cyberattacks that propagate across supply chain networks (Welburn et al., 2020, estimates supply chain linkages and the network effects of firm-level disruptions), and explore potential substitution effects as firms switch to uncompromised suppliers. Furthermore, given that cyber SCRM entails addressing risks to networks within networks, future research could model network structure deeper than interfirm networks and begin estimating the digital networks that support them. Using methods from data science that can leverage machine learning and artificial intelligence in conjunction with alternative large datasets, such as those on software prevalence, could enable a foray into further detailed analysis of cyber networks.

Strengthening the Manufacturing and Defense Industrial Base and Supply Chain Resiliency of the United States), and builds on the Annual Industrial Capabilities Report mandated by the Congress pursuant to section 2504 of title 10, United States Code. The report shall identify areas where civilian supply chains are dependent upon competitor nations, as determined by the Secretary of Defense.

Figure B.1. Discrepancy Between Supply Chain Perceptions and Reality

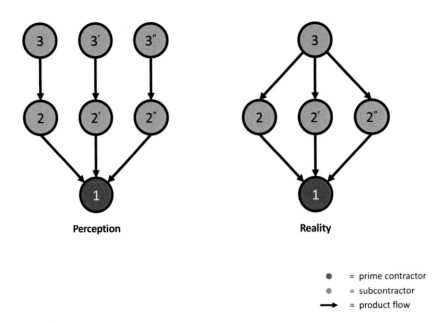

● = prime contractor
● = subcontractor
→ = product flow

NOTE: In this figure, the supply chain on the left represents a perception of many independent suppliers, and the supply chain on the right represents the reality of a single upstream source. The latter implies something closer but not equivalent to a single chain. Here, "1" could signify either a prime contractor or the ultimate customer.

Establishing Needs and Priorities for Supply Chain Response, Recovery, and Resilience

While establishing needs and priorities is a core DAF function, an evidence-based understanding of needs and priorities for response, recovery, and resilience to cyberattacks would better support policymaking and implementation. We suggest a three-part approach that would consider response, recovery, and resilience in the context of a supply chain's role in mission attainment. First, develop simple tools and mental models, possibly with a series of "what if?" questions to uncover possible mission-critical needs. Starting from the end of a cyber event, one could ask something like "what do we need to do to ensure that we have the parts we need to fly our mission, and what could prevent us from obtaining them?" Notwithstanding the importance we assign to consequences, we also recognize the value of triangulating with questions from the beginning of the event, e.g., "If X happens, why will it matter to acquisitions, and how will it affect our mission?" Other common approaches to uncovering such needs include gaming and TTXs, which are already widely used in deliberations on cyber-related risks. Second, look to the evidence on risk in each area to correlate perceived and actual needs. Framed in terms of a particular method, the research could develop something akin to a logic model and then validate the model empirically. Third, develop objective criteria and a consistent procedure to sort actual needs according to DAF priorities. This approach would involve taking a hard look at needs, separating true needs from wants, and setting priorities systematically.

Examining the Utility and Limits of Private-Sector Risk Reduction

DAF may look toward private-sector efforts to manage risks and find intriguing ideas, but differences between DAF and private-sector interests—as well as practical concerns about the data, methods, and supply chain visibility that would be needed to adequately support risk assessment—may limit the ability of would-be private-sector solutions to address DAF's concerns. For example, as we have noted in this report, industrial suppliers' incentives could lead them to underinvest in security relative to DAF or their own objectives. With such concerns in mind, we recommend pursuing research that examines the potential for improving the alignment of DAF interests and private-sector incentives; for cyber insurance to better contribute to response, recovery, and resilience; for strategic use of information; and for increasing the costs or difficulty of attacks.

What Steps Can Be Taken to Better Align Industry's Interests and DAF's Needs?

Inherent differences in public- and private-sector perspectives on security, response, and recovery, coupled with a tendency for businesses to underinvest, suggest the need for combinations of incentive mechanisms to help bring industry's interests into alignment with the DAF's needs. Research that further explores suppliers' incentives to invest—or not—as well as opportunities to alter their incentives, could contribute to policy development in this area. Our literature review suggests potential inroads at the nexus of game theory and network analysis and through explicit consideration of risks of disruption and exploitation. Insomuch as both types of risk interact with the configuration of a supply chain, research may point to some configurations that are better than others and opportunities for DAF to encourage industry to adopt the better configurations. Other promising research options include a deeper exploration of coordination among suppliers and how policy or contracting mechanisms could encourage suppliers to invest in the greater good and, perhaps, their own good. Along those lines, implementation of DoD's CMMC program, which had only just commenced at the time of our research, might provide a fruitful case study on incentives, behavior, and outcomes. Finally, in this vein, we suggest looking more closely at the implications of differences in IT and OT for cyber SCRM, including for technology refresh and policy compliance, possibly with capital replacement models that can incorporate the potentially interrelated effects of aging—or the passage of time—and risk.[59]

What Can Cyber Insurance Do for Response, Recovery, and Resilience?

If rates are tied to actions to mitigate risk, so as to avoid moral hazard and engender desirable behavior, insurance could reduce the risk of bad outcomes—or improve the timeliness of response and recovery—from DAF's perspective. However, rates must be targeted to the right

---

[59] For an example of an application of this type of model with and without uncertainty, see Greenfield and Persselin (2002). Greenfield and Persselin (2003) does not include the stochastic model but provides a fuller discussion of the underlying modeling method.

aspects of risk and priced appropriately, each of which presents difficulties of its own. Moreover, DAF would need some enforceable assurance that its suppliers will restore service to DAF first, but this might make sense only for a business for which DAF directly or indirectly constitutes a substantial share of the customer base. To start, we recommend careful consideration of the terms of policies and their requirements and benefits vis-à-vis security, response, and recovery, paying particular attention to DAF's interests, as a benchmarking and tracking exercise.[60] For example, do insurance policies offer discounts for good hygiene, or do they impose requirements for security measures prior to issuance? How do underwriters assess the differences in hygiene and measures so that discounts reflect actual conditions and behavior? In addition, do policies largely involve indemnification, or do they explicitly seek to facilitate response and recovery or build resilience? We also suggest looking for opportunities to formally model and empirically test the extent to which behavior and outcomes differ among insured and uninsured businesses. Research could also explore the difference between surmountable and insurmountable limitations on data and reporting on high-cost, low-frequency events.

What Part Can Information Play in Risk Reduction?

The literature on cyber risks is expansive, taken as a whole, but leaves us without clear guidance on when or how to share information and with whom, including among defense industrial suppliers. The results of our analysis suggest that information-sharing—either honestly or deceptively—can be "good" or "bad" for risk reduction, depending on the broader context and specific circumstances. In some instances, for example, suppliers might be able to create confusion among attackers and deter attacks; in others, however, sharing can increase an attack surface or create openings for exploitation. Research that extends or unifies game-theoretic and network models could help identify opportunities to use information more strategically and safely to reduce risk by more carefully distinguishing types and modes of sharing. To move beyond pure abstraction, the research would need to be tied to DAF-relevant empirical applications, e.g., regarding relationships with and among suppliers and needs to protect sensitive information.

What Can Businesses Do to Raise the Costs or Difficulty of Attacking?

Our brief consideration of the costs of attacking (Box 4.1) suggests a gulf between any interest in using a cost-based strategy to dissuade attackers and the knowledge of costs for doing so. Still, there might be opportunities to make attacks more difficult, hence "costlier," in terms of the Boolean attack model, for example, by identifying opportunities for accelerating hardware refresh in industry. A long-standing business paradigm has been to extend the useful life of hardware, thereby decreasing depreciation. However, that approach provides would-be

---

[60] Others, including OECD (2017) and Romanosky (2019), have considered these issues closely; given the evolving nature of the industry, new benchmarks and tracking could add value.

cyberattackers with enough time to both develop and exploit threats: the longer the useful life of hardware, the longer the useful life of a hardware exploit. The desirability of a target, however, might decrease with reductions in its useful life. A sufficiently short useful life might even deny the attacker the ability to develop successful and cost-effective exploits. We recommend research that explores the trade-offs between the costs associated with reduced useful life and reductions in the probability of compromise by a strategic cyberattacker and opportunities to shift the balance through policy mechanisms. Such research could, as suggested earlier in the context of aligning incentives, draw on standard approaches to capital replacement modeling.

It might also be possible to raise the costs or difficulty of attacking by training employees more effectively across the NIST functions. Given the prevalence of phishing, social engineering, etc., and the role of individual behavior in risk reduction, we see further research opportunities for employee training to avoid, respond to, and recover from cyberattacks. Advancements in training, perhaps through simulated environments and digital gaming, could enhance compliance, security, and response. DAF might look to best practices across military and nonmilitary environments for creative—and vetted—training options.

Crafting a Comprehensive Strategy for Cyber SCRM

Taking a more-comprehensive approach to developing a strategy for cyber SCRM would require better information about the form and extent of the trade-offs among risk-reduction objectives. That, in turn, would require improvements in risk assessment that may be unattainable. Given the potential for lasting uncertainty about risk parameters in the cyber domain, a comprehensive strategy for addressing cyber-related risk must itself also be resilient. Research in this area could involve testing proposed strategies for their robustness under varying conditions. For this purpose, we suggest considering a combination of thought models and tools for dealing with uncertainty in policymaking, e.g., drawing from Lempert et al.'s (2006) work on robustness. DAF doing something specific to stop one type of cyberattack might not help in other contexts and could make things worse in light of possible trade-offs among risk-reduction objectives. Approaches that are highly tailored may not be interchangeable, even with respect to wholly cyber concerns. Reaching out to consequences could sidestep some trade-offs because actions to improve resilience, such as those that create redundancy without shared vulnerabilities or introduce technological separation, might matter across the board for both *cyber* and *SCRM*.

That said, we would expect substantial gains from research that leads to a better understanding of the trade-offs among measures so that measures can be implemented that would come closer to optimizing across concerns. True optimization seems unlikely in this context, given the extent of the uncertainties. However, trying to balance concerns in light of potential trade-offs among objectives, with more insight into how big the trade-offs are, could represent a step closer to optimality. Finally, we recommend testing proposed approaches against priorities, possibly through gaming, TTXs, simulations, or pilot programs that can introduce specific elements of the policy in different combinations and all together.

Appendix C. Game-Theoretic Foundations

Research on game theory can help improve our understanding of the implications of the intentionality of threats that traditional risk assessment, which tends to focus on hazards, does not address.[61] In traditional risk assessment, efforts to mitigate the largest hazards in a system work toward reducing that system's overall risk, without concern for behavioral responses, but threats born of strategic interaction will shift as intelligent and strategic attackers find ways around risk-management strategies. In this appendix, we consider how some game-theoretic approaches—or *games*—can shed light on different concerns about how human behavior might affect the results of risk mitigation.

Attacker-Defender Games

Game-theoretic methods can provide analytical tools for understanding defensive strategy in the face of strategic threats. Following the foundational studies on games of strategy from Schelling (1960) and Dresher ([1961] 2007), the game-theoretic literature on security and defense has expanded considerably. Notably, this literature has grown to provide insight into policy options for modern security threats—from terrorism to cybersecurity—in which defensive decisions (i.e., protect, harden, detect, deter) alter the behavior of a strategic attacker. We do not attempt an exhaustive review of this expansive literature, but, instead, we draw out policy-relevant insights from an especially relevant subset.

The interdependent security problem studied by Kunreuther and Heal (2003) provides a sound starting point. Kunreuther and Heal (2003) introduce a model of strategic interaction between defenders with interdependent systems each confronted by a security investment problem. Importantly, each defender has only one system or potential target, and the risk depends on the actions of the other defender. For identical defenders, Kunreuther and Heal (2003) find two *Nash equilibria*—or stable outcomes—that represent extremes, in which it is either optimal for both or neither to invest in security.[62]

To understand the intuition behind the extremes, imagine a case in which two businesses must decide independently whether to invest in security, but if either business does not invest and then experiences an attack, both will bear the costs of the attack. This might occur if, for example, they exchange data on a common platform. If only one business invests in security and

---

[61] Here, we maintain the same distinctions between unintentional hazards and intentional threats that we made in Appendix A.

[62] A game in which players choose their strategy simultaneously is in Nash equilibrium when each player chooses their best response to the strategy of the other players and when no player could improve their payoffs by changing strategy.

if an attack occurs, the business that invested would end up bearing the costs of its investment and the attack and, with hindsight, would have been better off forgoing the costly but fruitless investment. Of course, both defenders would be even better off if they both invested in security and avoided the costs of being attacked. However, recognizing the possibility of incurring two costs, the defenders might choose not to do so.

The security challenge real-world decisionmakers face may be more complex in ways that would affect outcomes for security investments and, ultimately, for operations. For example, decisionmakers often face trade-offs associated with allocating resources across multiple possible targets. In that case, as Bier, Oliveros, and Samuelson (2007) demonstrates, the all-or-nothing result might not—or, likely, would not—hold.

Bier, Oliveros, and Samuelson (2007) addresses the inherent trade-off of allocating defensive resources to multiple vulnerable locations where defending one location will lead a strategic attacker to attack another location. This game is depicted by the sequence of player moves in Figure C.1, in which the vectors represent a range of possible options for acting. In the game, a defender chooses resources to allocate to protecting two sites, and an attacker chooses which site to attack. In the authors' construction, the defender allocates resources in a way that reduces the probability of success at a nonincreasing rate, where the resource allocation comes at a cost and attackers can only attack one location. Bier, Oliveros, and Samuelson (2007) solves the game both sequentially (defender moves first, and attacker moves second) and simultaneously (defender and attacker move at the same time), finding that the attacker's best response is the same in both cases; thus, any differences in the equilibrium result from the defender's preferences. In the sequential case, the defender has a *first mover advantage*, which is an inherent advantage that allows the player who moves first to set the game up to their advantage. In this case, it means that the defender can steer the attacker's attention according to its—the defender's—own preferences.

Figure C.1. Defender-Attacker Game Setup

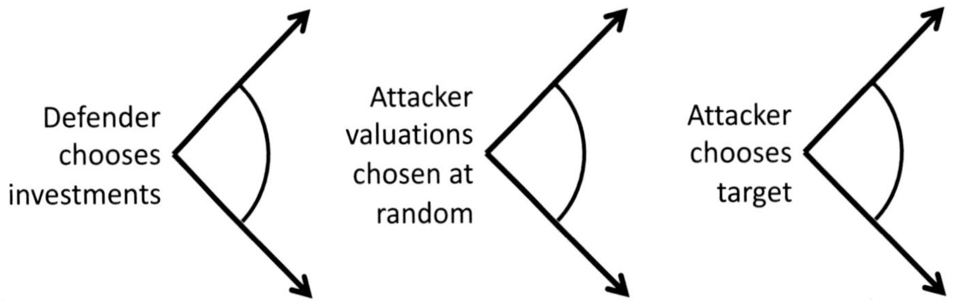

NOTE: This figure is adapted from Bier (2007) and shows the sequence of a basic defender-attacker game in the style of Bier, Oliveros, and Samuelson (2007), where the vectors suggest a range of possible options for acting.

The work of Bier, Oliveros, and Samuelson (2007) has general implications for defending against strategic attackers. It suggests the following:

- A defender with finite resources with multiple targets to protect faces a trade-off: Increasing protection at one location may increase the probability that an attack will be successful at another location. However, while an attack depends on the probability of success, it also depends on the attacker's valuation. Thus, in equilibrium, a defender may choose to leave one location undefended while protecting another if the value of the undefended target is sufficiently low from the perspective of either the defender or the attacker.
- Given the potential for an attack at one location to negatively affect other locations, centralized defensive allocations that consider all impacts are preferable.
- The defender's optimal strategy may differ according to the attacker's opportunity costs, insomuch as a defender may be better off with light defenses when the attacker's opportunity cost, perhaps posed by an attractive alternative target or a large diversion of resources for the required attack, is high.
- There are gains from making defensive allocations public when the defender moves first and the attacker can fully observe its moves, implying that some information-sharing can be beneficial.

Bier (2007) and Bier, Oliveros, and Samuelson (2007) note that, as the number of vulnerable locations grows, the defender can cost-effectively reduce the probability of a successful attack only if the number of locations the defender values is bounded (i.e., finite). This finding, while general and seemingly only theoretical, has implications for managing cyber risks to supply chains if, for example, the number of locations in a supply chain is finite but unknown, which might be the case in a multitiered and complex industry, such as the defense industry (see Chapter 4).

A broader literature has followed from Bier, Oliveros, and Samuelson (2007), often focusing on real-world complexities that are hard to capture in highly stylized models. For example, Hausken (2008) introduces network-like concepts and addresses the implications of interdependent systems for reliability. According to Hausken, when a system is configured in series (Figure C.2), the attacker benefits because it would only need to attack one node in the chain to cause a disruption. However, when a system is configured in parallel and is truly distinct, the defender benefits because the attacker would have to attack each node separately. This finding suggests that redundancy in the cybersecurity context can hold value but that the technology must be separate and, preferably, distinct. Duplicative redundancy could increase rather than decrease risk, by expanding the attack surface.

Zhuang, Bier, and Alagoz (2010) extends the literature further, by considering the implications of imperfect information, signaling, and repeated interaction. The authors present a game with a defender and an attacker, but, in this case, the defender has private information with respect to its characteristics or type, meaning it knows something about itself, such as whether it has invested in security, that the attacker does not know, unless the defender provides a signal. The defender can choose to signal its type truthfully, can choose to signal deceptively (i.e., mix true and false signals), or can choose to maintain secrecy. Ultimately, the authors found that a defensive strategy of secrecy and deception, applied over time, can prove to be a more cost-effective security strategy than candor. Thus, they point to the potential benefits of limiting disclosures and laying false trails as security measures, suggesting that sharing less information or sharing it deceptively could be better than sharing more information and sharing it honestly in some circumstances.

Figure C.2. Connections in Series Versus in Parallel

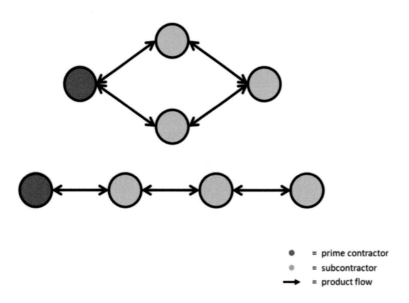

SOURCE: Author's illustration, based on Hausken (2008).
NOTE: This figure shows a system connected with nodes in series (bottom) and in parallel (top).

Haphuriwat, Bier, and Willis (2011); Jackson and LaTourrette (2015); and others have used game-theoretic approaches to model strategies for defending against strategic attackers who engage in terrorism. For example, Haphuriwat, Bier, and Willis (2011) addresses the concern of terrorists' abilities to smuggle nuclear bombs into the United States in freight containers. The authors found that, if defenders can credibly signal a high cost of retaliation, partially inspecting freight containers for nuclear bombs can be sufficient to deter attacks, a finding with broader implications when detection is difficult. Similar approaches in the use of game-theoretic models to understand defensive investment in response to terrorist threats have been used in aviation security (e.g., Jackson and LaTourrette, 2015) and road networks (e.g., Bell et al., 2008), finding advantages in a layered security approach. Consider the example of airport security: While visible defenses may exist at security checkpoints, at entryways, and elsewhere in an airport, layered security may start with ticket purchases and no-fly lists, may continue with in-airport behavioral monitoring, and may include in-flight security with air marshals.

Games in Cybersecurity

Notably, the method of studying strategic interactions between defenders and attackers has also been applied to a subject of this report, cybersecurity. We have not attempted to provide an exhaustive review of the literature applying game-theoretic approaches to cybersecurity; for that, see Roy et al. (2010) and Do et al. (2017) for early and recent surveys. Instead, this section highlights relevant and notable papers on information-sharing, deception, deterrence, and supply chain risk.

Hausken (2007) studies the benefits of information-sharing (e.g., disclosure of threats, vulnerabilities, and attacks) across firms. The author found that firms tend to underinvest and free ride on the efforts of others. Hausken also found that the most efficient defense is achieved when a social planner has the first-mover advantage and controls information-sharing. This is particularly true when high defense efficiencies can deter attacks entirely. Hausken's findings provide additional evidence that coordination (perhaps through a single well-respected social planner) can enhance security and have a deterrent effect. Nagurney and Shukla (2017) presents a multifirm model of cybersecurity investment in competitive and cooperative environments. Similar to Hausken (2007), Nagurney and Shukla found gains from information-sharing among firms and quantified its monetary and security benefits.

A growing literature has sought to understand strategies of deterrence applicable to cyberspace. Edwards et al. (2017) presents a game between an attacker and a defender specifically focused on the policy question of attribution, finding that when attribution confidence is high, the defender gains deterrence value from publicly attributing blame to the attacker. Practically speaking, attribution is fundamental to the use—or threat of use—of retaliatory measures to deter adversaries in cyberspace; simply put, if you cannot accurately place blame on an adversary, you cannot retaliate or credibly threaten retaliation against it, at least not as a targeted matter.[63]

Baliga, de Mesquita, and Wolitzky (2020) presents a model with a single defender and multiple attackers in which attackers can each chose to attack. When attacked, the defender receives an uncertain signal of attribution and faces the choice of whether to retaliate. The authors found an endogenous complementarity among attackers, wherein the most aggressive attackers increasing aggression can lead all others to increase their levels of aggression. This finding is based in the uncertainty of attribution in which the more frequently an attacker attacks the defender, the more likely it is that the defender is going to catch that attacker and attribute attacks to it. By analogy, if the highway patrol is most likely to see and pull over the fastest driver on the highway, the other drivers will know this and pace themselves accordingly, driving just slower than the fastest car to avoid getting a ticket. Consequently, the faster the lead driver goes, the faster the others are likely to drive. Similarly, the authors found that, for cyberspace,

---

[63] Inability to attribute does not, however, rule out the possibility of punishing all or many for the acts of one.

when the most aggressive attacker is most likely to get caught and punished, any other attackers have the incentive to increase their aggression, just as long as they stay behind the most aggressive one. This research suggests that weak attribution can promote aggression, insomuch as attackers will tend to ratchet up their aggression to nearly match that of the most aggressive attacker.

Welburn, Grana, and Schwindt (2023) defines a basic attribution game between an attacker and a defender, similar to that in Edwards et al. (2017), in which attackers and defenders choose whether to attack or retaliate, respectively, but defenders must choose without knowing whether they have been attacked. Welburn, Grana, and Schwindt built on the game by adding *signaling*. In their model, the defender has private information with respect to its capability to retaliate (high or low), which it can costlessly signal truthfully or not truthfully to the attacker. The authors contributed further evidence of the value of deception by finding that it is never in the best interest of the defender to signal perfectly. They also found that successful deterrence, while potentially feasible, is likely to be adversary specific, depending on both the defender's capability to retaliate against the adversary and the adversary's perceived penalty from retaliation. When the probability of successful attribution and the capability to retaliate are both high, the defender can find success in cyber deterrence. In some instances, where the defender has an especially high capability to retaliate, Welburn, Grana, and Schwindt found that, by inducing an attack, the defender may actually improve its own payoffs through a strong public retaliation with high confidence. This result, while seemingly counterintuitive, amounts to a public show of retaliatory capability strengthening deterrence posture for other would-be attackers.

Game-Theoretic Models of Cyber Risks to Supply Chains

A few recent papers have applied game theory to the problem of cyber-related risks to supply chains and, thus, touch on network-related concerns.

Bandyopadhyay, Jacob, and Raghunathan (2010) models cybersecurity for two firms that are connected through a communication network in a supply chain with varying degrees of integration. An *unintegrated* supply chain is one in which the two firms are connected only through shared communication, and an *integrated* supply chain is one in which they are connected through shared communication and, to some extent, shared production, ownership, financing, and decisionmaking. In this model, each firm can be attacked directly or indirectly, through a breach that propagates through the communications network, and can invest in its own security to decrease the probability of a successful attack. The authors found that unintegrated firms decrease investment—in effect, behaving like free riders—as network vulnerability increases, that loosely integrated firms also decrease investment as shared vulnerabilities increase, and that tightly integrated firms increase investment when the vulnerabilities increase. Their results imply that the extent of integration—above or below a threshold value—can determine whether firms choose to increase investment or free ride in response to greater vulnerability.

These results suggest that firms may underinvest in security when supply chain integration is low, despite shared vulnerabilities across communication networks, and may continue to underinvest even with a modicum of integration. Thus, Bandyopadhyay, Jacob, and Raghunathan (2010) points to the risks of shared digital networks and suggests a potential role for coordinated investment strategies among firms.[64]

Nagurney, Nagurney, and Shukla (2015) presents a supply chain network model between customers that are retailers and a single tier of suppliers. While the focus is on the effect of competition among the retailers on security investment, the authors found evidence that increasing interdependence among supply chain connections may increase susceptibility to cyberattack, mostly because of the serial connection of supply chain nodes.

Simon and Omar (2020), which comes the closest in this literature to addressing DAF's concerns, models the strategic interaction between a single cyberattacker, attempting to compromise a supply chain, and a separate defender at each node or firm in the supply chain, choosing to protect the node. The authors constructed a model wherein each node in an interconnected supply chain is vulnerable to disruption by cyberattack. A successful attack causes node-dependent levels of damage, where an attack on more critical nodes causes more damage and leads to the assumption that damage increases as nodes become closer within the supply chain. For both the attacker and the defender, the authors model the actions in two ways: The attacker is represented as either *nonstrategic* (they attack nodes at random) or *strategic* (they choose their attack strategy according to the strategy of the defender), and defenders are represented as either *coordinated* or *uncoordinated* with regard to their decisions about security investments. In both cases and for both types of attacker, the study finds suboptimal investment in security without coordination.

While Simon and Omar (2020) does not address policy implications, some such implications can be inferred for DAF from its analysis and discussion. For example, it notes that a lack of coordination among firms within a supply chain can lead to poorly allocated investment and undermine risk management but suggests that, in the absence of formal coordination, it may be in the interest of larger firms to subsidize the cybersecurity investments of smaller firms. Were these cross-subsidies to occur, the private sector might be inclined to address at least some potential gaps in security on its own, without aid from policy. Alternatively, against a strategic attacker, increasing the investment at larger firms is best and would push the attacker toward target indifference, thereby spreading out risk more evenly. Moreover, the results of the article's excursion on indirect damage show that a lack of coordination across the supply chain can lead

---

[64] Useful extensions include considering the effects of *N* firms rather than just two firms and capturing the impact of risk aversion on investment decisions. The extension to *N* firms overlaps with the results of Nagurney and Shukla (2017) and Baliga, de Mesquita, and Wolitzky (2020).

to underinvestment, particularly for suppliers immediately upstream, meaning closest to the consumer, as in the case of a DAF prime supplier. For DAF, this implies a potential benefit from coordination among prime suppliers and their input suppliers or subcontractors.

Appendix D. Insights from Network Analysis

In this appendix, we formalize the discussion of interactions between cyber-related risks and supply chains and calculate the risks of disruption and exploitation for each case that we described in Chapter 5. Furthermore, we review approaches to studying production networks that might bear on SCRM and then take a closer look at the use of network analysis to understand cyberattacks on supply chains.[65] The approaches build on the seminal work of Leontief (1966) to study the impact of economic networks, specifically IO networks, in propagating sector- or firm-specific risks across network connections.

Analysis of Cyber Threats to Stylized Supply Chains

For the specific question of how adding a node to the supply chain adds to *overall risk*, defined as the probability of an event multiplied by a measure of the severity or impact, we can find answers by attaching probabilities to simple representations of supply chains. That is, given the probabilities of cyberattacks on nodes throughout the supply chain, we can estimate the overall probabilities of damage from cyber-related threats for the entire supply chain. Assuming that all nodes have the same criticality and the same ultimate effect on the supply of the good, i.e., have the same impact, the estimated probabilities would provide an approximate measure of overall risk. We differentiate between risks of supply chain *disruption* and *exploitation*. The former can result in a loss of product availability or quality or an increase on cost; the latter, consisting of infiltration and exfiltration, could result in a loss of product integrity or information. In this section, we estimate the risks of disruption and exploitation for stylized representations of supply chains to draw general insight into the nature of cyber supply chain risk and potential mitigations.

We start by estimating the probability of disruption for the three stylized supply chains shown in Figure D.1, as Cases A, B, and C. In Case A, disruption can come from an attack on either the single prime supplier (node 2) or the customer (node 1). The same is true for Case C, where an attack on a second-tier supplier (node 3) adds an additional opportunity for disruption. Case B also adds a node, relative to Case A, but instead of lengthening the chain, it includes two

---

[65] In economics, the study of production networks is closely related to the study of supply chains. Here, we review some especially relevant work on production networks; for further reading, Carvalho and Tahbaz-Salehi (2019) provides a useful review of the literature and general findings.

substitute prime suppliers (nodes 2 and 2′), which implies that a disruption to this tier would require an attack on *both* nodes 2 and 2′.[66]

Figure D.1. Three Basic Stylized Supply Chain Cases

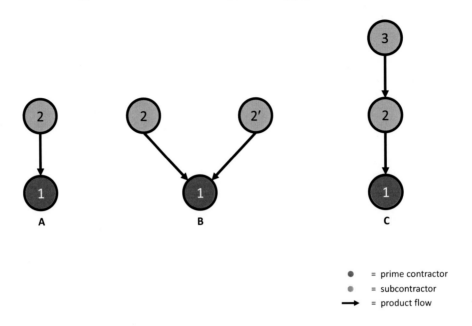

● = prime contractor
● = subcontractor
→ = product flow

NOTE: This figure displays three basic stylized supply chains: In supply chain A, supplier 2 supplies inputs to supplier 1; in supply chain B, suppliers 2 and 2′ supply substitute inputs to supplier 1; and, in supply chain C, supplier 3 supplies inputs to supplier 2, which, in turn, supplies inputs to supplier 1.

We can formally estimate the risk of disruption in each of the three examples shown in Figure D.1. Define D as the probability of disruption of the overall supply chain where π_i is the probability of disruption of a tier and p_{ji} is the probability of attack on a node j in tier i. Then, for Case A, left of Figure D.1, the probability of disruption, D_A, is estimated as follows:

$$D_A = \pi_1 + \pi_2 - \pi_1\pi_2 = 0.5 + 0.5 - 0.25 = 0.75. \tag{1}$$

Similarly, for Case B in Figure D.1, the probability of disruption, D_B, is

$$D_B = \pi_1 + p_{22}p_{2'2} - \pi_1 p_{22}p_{2'2} = 0.5 + 0.25 - 0.125 = 0.625. \tag{2}$$

Note that, if we define the probability of disruption to tier 2 as $\pi_2 \equiv p_{22}p_{2'2} = 0.25$, then D_B is more simply stated as follows:

$$D_B = \pi_1 + \pi_2 - \pi_1\pi_2 = 0.5 + 0.25 - 0.125 = 0.625. \tag{3}$$

Finally, for Case C, the probability of disruption, D_C, is estimated as follows:

---

[66] As a practical matter, this could mean that the consumer purchases all goods from one supplier or the other, but can switch suppliers instantaneously, or that it purchases some from each, but could immediately expand its purchases from one or the other to meet its needs.

$$D_C = \pi_1 + \pi_2 + \pi_3 - (\pi_1\pi_2 + \pi_2\pi_3 + \pi_1\pi_3 - \pi_1\pi_2\pi_3)$$
$$= 0.5 + 0.5 + 0.5 - (0.25 + 0.25 + 0.25 - 0.125) = 0.875. \tag{4}$$

The probability of disruption, D, generalizes to

$$D = \sum_{i=1}^{n} \pi_i - \sum_{m=1}^{n} \prod_{i=0}^{m-1} \pi_i, \tag{5}$$

where

$$\pi_i = \prod_{j=0}^{m} p_{ji} \; \forall \, i. \tag{6}$$

Thus, adding tiers to lengthen a chain increases the probability of disruption, while adding substitute nodes within a tier reduces the probability of disruption.

Next, we consider risks of exploitation, but, for simplicity and as a point of contrast, we focus on exfiltration. If all nodes contain the same information, an attacker who seeks to extract information from the supply chain would only require access to one node:[67]

$$E = 1 - \prod_{j=0}^{m}(1 - p_{ji}). \tag{7}$$

Using the same three examples as in Figure D.1, the probability of exfiltration for Cases A, B, and C (E_A, E_B, and E_C respectively) can be found as follows:

$$E_A = 1 - (0.5 \times 0.5) = 0.75$$

$$E_B = 1 - (0.5 \times 0.5 \times 0.5) = 0.875$$

$$E_C = 1 - (0.5 \times 0.5 \times 0.5) = 0.875. \tag{8}$$

Relative to the two-tier supply chain (Case A), adding a redundant node (Case B) and another tier (Case C) increases the risk of exfiltration. Simply, each additional node expands the attack surface, by the same amount, and increases the risk of exfiltration, also by the same amount.

Furthermore, we can introduce additional nuance by including information loss with each tier. As information is shared further upstream, e.g., at nodes in the third tier, as compared with the second, its exfiltration may become less harmful to the ultimate customer if lost. With some equivalence, exfiltration of customer data may also be less likely at higher tiers in the supply chain. Here, we assume that information decays at rate δ ($0 < \delta < 1$) with each connection, which results in the following probability of exfiltration:

$$E = 1 - \prod_{j=1}^{m} \delta^i (1 - p_{ji}). \tag{9}$$

Then, if the impact of exfiltration is equal for all nodes, the marginal risk of adding a tier decreases with each added tier, which could imply that a supplier at a higher tier needs less

---

[67] For purposes of infiltration, an attacker might still need to reach all nodes.

protection than a supplier at a lower tier. This points to a value of adding redundant nodes at tiers further upstream.[68] This benefit can be illustrated through a simple example.

Consider the two stylized two-tier supply chains shown in Figure D.2. Both introduce redundancy through the inclusion of a single substitute supplier. However, on the left, redundancy is introduced in the third tier; on the right, redundancy is introduced in the second tier.

Figure D.2. Supplier Redundancy in Stylized Supply Chains

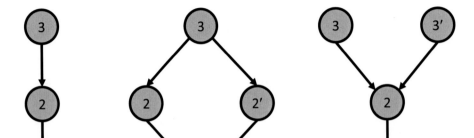

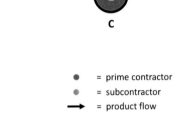

NOTE: This figure displays three ways of introducing redundancy in a three-tier supply chain. Supply chain A presents a baseline supply chain with just three suppliers; supply chain B introduces a redundant supplier in the second tier; and supply chain C introduces a redundant supplier in the third tier.

Using the same approach as previously, we next estimate the risk of disruption and exfiltration for the two examples shown in Figure D.2. For simplicity, assume that all probabilities, $\pi_i = 0.5$, are equal. Then we can show that the risk of disruption for Cases B and C are equal. Note that, in Case B, $\pi_2 \equiv p_2 p_{2'}$ and, in Case C, $\pi_3 \equiv p_3 p_{3'}$. Thus, estimating the two probabilities of disruption for Case B,

$$D_B = \pi_1 + \pi_2 + \pi_3 - (\pi_1 \pi_2 + \pi_2 \pi_3 + \pi_1 \pi_3 - \pi_1 \pi_2 \pi_3)$$

---

[68] If, however, one imagines an alternative scenario in which information grows in value upstream rather than decaying, the opposite can be true. It might, for example, be the case that an upstream supplier has essential information about the design of a subcomponent that cannot be obtained elsewhere. Specifically, information growth ($1 < \delta < \infty$) implies that a supplier at a *lower* tier needs less protection than a supplier at a *higher* tier. This points to a value of adding redundant nodes at tiers closer downstream.

$$D_B = \frac{1}{2} + \frac{1}{4} + \frac{1}{2} - \left(\frac{1}{8} + \frac{1}{8} + \frac{1}{4} - \frac{1}{16}\right) = \frac{13}{16} = 0.8125. \tag{10}$$

And Case C,

$$D_C = \pi_1 + \pi_2 + \pi_3 - (\pi_1\pi_2 + \pi_2\pi_3 + \pi_1\pi_3 - \pi_1\pi_2\pi_3)$$

$$D_C = \frac{1}{2} + \frac{1}{2} + \frac{1}{4} - \left(\frac{1}{4} + \frac{1}{8} + \frac{1}{8} - \frac{1}{16}\right) = \frac{13}{16} = 0.8125, \tag{11}$$

the case of the simplest three-tier strand, which is $D_A = 0.875$.

The risk of exfiltration, however, is different across the three examples shown in Figure D.2. For Case B, the probability, hence risk, of exfiltration is

$$E_B = 1 - \prod_{j=0}^{m} \delta^i\left(1 - p_{ji}\right) = 1 - \left(\frac{\delta}{2}\right)\left(\frac{\delta^2}{2}\right)\left(\frac{\delta^2}{2}\right)\left(\frac{\delta^3}{2}\right) = 1 - \frac{\delta\delta^2\delta^2\delta^3}{16}, \tag{12}$$

while, for Case C, the probability of exfiltration is

$$E_C = 1 - \prod_{j=0}^{m} \delta^i\left(1 - p_{ji}\right) = 1 - \left(\frac{\delta}{2}\right)\left(\frac{\delta^2}{2}\right)\left(\frac{\delta^3}{2}\right)\left(\frac{\delta^3}{2}\right) = 1 - \frac{\delta\delta^2\delta^3\delta^3}{16}. \tag{13}$$

Since $\delta < 1$, $\delta\delta^2\delta^2\delta^3 > \delta\delta^2\delta^3\delta^3$ implies that $E_C < E_B$. Therefore, in this example, adding a node in the third tier may lead to lower risk of exfiltration than adding a node in the second tier while resulting in the same level of risk to disruption.

Importantly, a lack of independence among nodes significantly alters these results. Consider the simple examples of the two-tier supply chains shown in Figure D.3. Our calculations in equations (1), (3), and (8) show that, in comparison with Case A, the example with redundant suppliers in the second tier (Case B) has a lower risk of disruption but higher risk of exfiltration. However, if the second-tier nodes 2 and 2′ are not independent (Case C) and share a common disruption-allowing vulnerability (i.e., a common cyber floodplain) that could be exploited to attack both with the same attack, then this example has both a higher risk of disruption and a higher risk of exfiltration.

Figure D.3. Supplier Independence and Interdependence

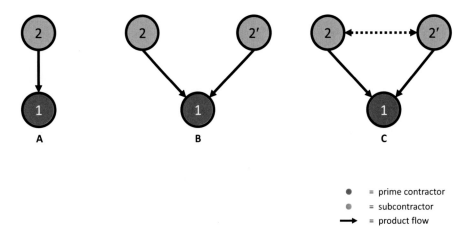

● = prime contractor
● = subcontractor
→ = product flow

NOTE: This figure displays three stylized two-tier supply chains. In supply chain A, supplier 2 supplies inputs to supplier 1; in supply chain B, suppliers 2 and 2′ provide substitute inputs to supplier 1; and, in supply chain C, suppliers 2 and 2′ are also dependent on common software, shown with a dashed line.

Therefore, using the stylized representations of supply chains, we can draw the following four general insights:

1. All things equal, adding a tier to the supply chain may add to risks of both disruption and exploitation, suggesting that risk mounts with deeper supply chains, especially if, as findings from game theory suggest, the length of the chain is uncertain.
2. All things equal, adding a redundant supplier to a single tier may decrease the risk of disruption while increasing the risk of exploitation, suggesting the possibility of trade-offs among risk-reduction objectives and options.
3. All things equal, adding a redundant supplier upstream rather than downstream can add less risk of exploitation if information decays with successive tiers, suggesting that, although such risk mounts with supply chain depth, it may do so at a decreasing rate.
4. All things equal, adding a redundant supplier reduces disruption risk only if the probability of a successful attack on the redundant supplier is independent of the probability for others. If it is not independent, however, this redundancy can add to both disruption and exploitation risk. Thus, common technological vulnerabilities among suppliers, such as through shared systems or like platforms, can undermine any benefits of adding suppliers.

Literature on Production Networks

In this section, we briefly review work on production networks, including some recent efforts to study cyber-driven shocks to production networks. That ties into concerns for what we refer to as the problem of "network attacks on networks" and the role of network structure. This literature works with a method known as *IO analysis*, because it can be used to trace the relationships between and among businesses, sectors, and economies through their use of inputs and production of outputs, in which the outputs of one constitute the inputs of another. While the literature using IO analysis is expansive, a subset of the literature studying disasters, both naturally occurring and maliciously induced, and economic shocks aptly demonstrates the salience of interdependencies.

For example, Santos and Haimes (2004) uses an IO model to examine the extent to which the economic impacts of a terrorist attack that reduces airline demand can propagate to other sectors. The authors found that the disaster risk of one sector, in this case air transportation, can create significant risks for others through connections among sectors. In the terrorist attack scenario, the authors estimated that large losses would also be incurred by hospitality services (travel arrangements, sightseeing), airline parts manufacturing, and oil and gas extraction, implying repercussions not just at the point of an attack but potentially across industries and up and down supply chains.

Santos, Haimes, and Lian (2007) extends that approach to consider the economic impact of a cyberattack on supervisory control and data acquisition systems. This analysis provides a framework for linking physical and economic impacts of cybersecurity scenarios that yields aggregate metrics for measuring risk and assessing security measures in terms of cybersecurity costs, network vulnerability, equipment downtime, and production delays, as well as estimates of macroeconomic effects. That is, the authors have provided a framework for policy analysis that traces how cyberattacks lead to ripple effects through sectoral and regional interdependencies. They applied the framework to consider the case of a cyberattack causing a disruption in U.S. Gulf Coast oil output, finding indirect sectoral and regional impacts.

Without regard to a particular form of hazard or threat, Acemoglu et al. (2012) provides a general mathematical framework for understanding how isolated shocks can pose large aggregate risks when network effects amplify their effects. Building on the IO linkages introduced by Leontief (1966), Acemoglu et al. found that idiosyncratic shocks on the microeconomy can propagate across economic networks, both upstream and downstream, leading to aggregate fluctuations in the macroeconomy. Furthermore, using a similar approach to study the susceptibility of financial networks to shocks, Acemoglu, Ozdaglar, and Tahbaz-Salehi (2015) found that the impact of shocks to networks depends on the structure of the underlying network. More specifically, the authors demonstrated that, while "ring networks" are the least resilient and that "complete networks" are the most resilient to small shocks, complete networks are the least resilient, and ring networks are the most resilient for large shocks (see Figure D.4).

Figure D.4. Ring Networks Versus Complete Networks

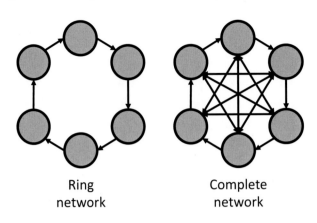

Ring
network

Complete
network

Others have looked more specifically at the role of firm-level production networks and the propagation of disruptions to individual firms and their customers. For example, Barrot and Sauvagnat (2016) constructed a dataset of interfirm network connections to study the propagation of shocks through supply chain relationships following natural disasters. They found that, following disruptions, businesses pass a significant share of their losses on to their customers, which, through further propagation through supply chains, can lead to significant economic loss. Similarly, Boehm, Flaaen, and Pandalai-Nayar (2019) studied the role of supply chain linkages in propagating natural disasters, specifically focusing on the propagation of disruptions following the 2011 Great East Japan Earthquake. In addition to providing consistent, reinforcing findings on propagation, Boehm, Flaaen, and Pandalai-Nayar (2019) reported that, following disruptions, firms drew nearly zero recontracts. The results imply that the firms were unable to respond to the shock by finding new suppliers. MacKenzie, Barker, and Santos (2014), Carvalho et al. (2020), and others have also studied the earthquake and found that a significant amount of economic losses came from the propagation of disruptions through supply chain networks.

Welburn and Strong (2022) extends this literature, focusing on the potential systemic risks of cyberattacks. The authors defined systemic cyber-related risks as those resulting from cascading failures, which ripple though digital networks and supply chains; common cause failures, which result from shared vulnerabilities; and large-scale independent cyber failures. Using an IO framework that is similar to those of Santos, Haimes, and Lian (2007) and Acemoglu et al. (2012), Welburn and Strong (2022) models the cascading effect as a cyberattack leads to losses that spread upstream and downstream through supply chains. The theoretical framework is applied to the case of single-day outages across five large firms, in which the authors found the potential for large aggregate effects from relatively small cyberattacks on key firms that propagate to firms both up and downstream in their supply chains. Furthermore, Crosignani, Macchiavelli, and Silva (2021) uses data on firm linkages to identify the propagation of disruptions within the supply chain networks following the 2017 NotPetya cyberattack. In a different approach (theoretical versus empirical modeling), both Welburn and Strong (2022) and Crosignani, Macchiavelli, and Silva (2021) find that the vast majority of supply chain damage following a cyberattack comes from propagations to downstream customers within the network. Furthermore, while firms were not able to substitute suppliers for new ones over the near term, Crosignani, Macchiavelli, and Silva (2021) finds that NotPetya led to adjustments in customer and supplier relationships in the long term, lasting years.

Modeling Foundations for Analysis of Production Networks

Following recent efforts to study firm-level productions networks (e.g., Welburn et al., 2020), we can define the foundations for the analysis of cyber-related risks to supply chain networks. Supply chains are each a partition of an overall economy made up of a set of n firms and a network of interfirm connections. This set of network connections is defined by both the adjacency matrix

$$\mathbf{A} = \llbracket a_{ij} \rrbracket_{i,j} \in \{0,1\}^{n \times n},$$

where $a_{ij} = 1$ if i supplies to j and 0 otherwise, and the weighted-adjacency matrix

$$\mathbf{W} = \llbracket w_{ij} \rrbracket_{i,j} \in [0,1]^{n \times n},$$

where w_{ij} is the share of i's output used as inputs by j. Thus, given a firm's output x_i, upstream flows x_{ij} and downstream flows x_{ji} are given as follows:

$$x_{ij} = w_{ij}x_i, x_{ji} = w_{ji}x_j, i,j = 1, \dots n. \tag{14}$$

Using this notation, we can define firm i's tier 1 suppliers as those for which $a_{ji} = 1$, tier 2 suppliers as those for which $a_{jk}a_{ki} = 1$, tier 3 suppliers as those for which $a_{jk}a_{km}a_{mi} = 1$, and so on.

Next, we define $\mathbf{x} = [x_1 \quad \cdots \quad x_n] \in \mathbb{R}^n$ as the vector of outputs and $\mathbf{d} = [d_1 \quad \cdots \quad d_n] \in \mathbb{R}^n$ as the vector of final demands by all firms. Using standard IO analysis, firm level output is defined as

$$\mathbf{x} = \mathbf{Wx} + \mathbf{d} \tag{15}$$

or

$$\mathbf{x} = (\mathbf{I} - \mathbf{W})^{-1}\mathbf{d} = \mathbf{Ld}, \tag{16}$$

where $\mathbf{I}$ is the identity matrix, and $\mathbf{L}$ is defined as the inverse Leontief matrix. Many have used this basic setting to describe how network interdependencies can propagate adverse events, which may start in a given sector or given firm, causing aggregate losses across supply chains and the broader economy.

A digital supply chain adds another form of interconnectivity. In exchange for goods and services, firm j provides digital information to firm i. We define the flow of digital information as d_{ji}, where the matrix

$$\mathbf{D} = \llbracket d_{ij} \rrbracket_{i,j} \in \{0,1\}^{n \times n}$$

represents a network of digital connections. Thus, for a given supply chain connection in which goods and services flow from i to j, there is also a digital connection in which information flows from j to i (i.e., $a_{ij} = 1 \Rightarrow d_{ji} = \delta$ or $\mathbf{D} = \delta \mathbf{A}^T$ where $\delta \in [0,1]$ a data transmission rate).

Using these network structures, we can define two broad types of cyberattacks: disruption and exploitation. *Disruption* represents an attack that propagates through the supply chain network $\mathbf{A}$ to reduce the ultimate quality of goods and services, while *exploitation* represents a data breach on the digital network $\mathbf{D}$. Define disruptive attacks on a given firm i as $\epsilon_i = 1$ for an attack and $\epsilon_i = 0$ otherwise, such that $\boldsymbol{\epsilon} = [\epsilon_1 \quad \cdots \quad \epsilon_n] \in \{0,1\}^n$. Similarly, define exploitation at a given firm i as $\phi_i = 1$ for a breach and $\phi_i = 0$ otherwise, such that $\boldsymbol{\phi} = [\phi_1 \quad \cdots \quad \phi_n] \in \{0,1\}^n$. Furthermore, define a defense vector $\boldsymbol{\pi} = [\pi_1 \quad \cdots \quad \pi_n] \in \{0,1\}^n$, such that π_i, the probability that attempted attack or breach is successful at a given node i, decreases with defensive investment.

For cyberattacks occurring such that $\epsilon_i \sim f(\mu, \sigma)$ and $\phi_i \sim f(\mu', \sigma')$, it can be shown that the cost of upstream disruption propagation is $\mathbf{L}\boldsymbol{\pi}'\boldsymbol{\epsilon}$ and the cost of upstream exploitation propagation is $\mathbf{L}\boldsymbol{\pi}'\boldsymbol{\phi}$. What could be shown is the role of network structure on cyberattacks. This could consider four cases:

1. a completely unprotected network, $\pi_i = 1$
2. a partially protected network, e.g., $\pi_i = 0.5$
3. a fully protected subnetwork, e.g.,

$$\pi_i = \begin{cases} 0.1, i \in \{1, k\} \\ 1, i \in \{k+1, n\} \end{cases}$$

4. optimal protection, $\pi_i = \pi_i^*$, where π_i^* is found by the solution to a linear program.

Each of these cases is equivalent to a game-theoretic model with a nonstrategic attacker with policy analogs. Case 1 considers policy options that cannot change defensive strategies but can change network structure (long versus short supply chains, ring versus hub-and-spoke supply chains). Case 2 considers the benefit of improving security for all. Case 3 considers the case in which only part of the supply chain must adhere to a higher security standard. Case 4 considers the case in which security investments can be targeted. There is interaction in all cases and the network structure. For example, Case 3 may lead to more improvements with a hub-and-spoke structure than ring (or vice versa).

Appendix E. Risk Management Methods

In this appendix, we walk through the different methods that informed our approach to characterizing cyber-related risks, including the threat environment. We start with a time-honored risk management framework that has served across the national security and other policy communities, then turn to a complementary cyberattack model based on Boolean logic. We have chosen to work with this risk management framework rather than the framework presented in Air Force Instruction 17-101 (2020) or the NIST cybersecurity framework (NIST, 2018b) because of its broader use and familiarity outside the cyber community. Finally, we provide a crosswalk between the risk management framework and the cybersecurity functions in NIST's cybersecurity framework because we draw concepts and vocabulary from the latter, such as *response* and *recovery*.

Risk Management Framework

Military guidance provides a five-step risk management process and accompanying risk assessment matrix that suggest the importance of both understanding risk and iterating cyclically toward solutions, perhaps indefinitely.[69] (See Figures E.1 and E.2.) Steps 1 and 2 of the five-step process involve risk characterization, while Steps 3, 4, and 5 pertain to decisionmaking and actions.

As discussed in Greenfield and Camm (2005, p. 49), which explores the framework,

> [r]isk control, which occurs in steps three, four, and five as part of risk mitigation, would involve developing a strategy for eliminating, reducing, or coping with the possibility of a hazard. By implication, the goal of risk mitigation is not necessarily risk elimination. In some instances, it may be preferable to accept some amount of "residual risk" and develop a response and recovery plan. . . . step three [of the process], "Develop [Controls] and Make Risk Decisions," also requires evaluation of controls for suitability, feasibility, and acceptability, where acceptability refers, in part, to cost-benefit assessment.

Department of the Army Pamphlet 385-30 (2014, pp. 9–10) and AFPAM 90-803 (2017, pp. 27–36) reaffirm this approach, explicitly or implicitly. In neither case does developing controls or making risk decisions equate to eliminating—or defending away—risk.[70] A *control* could reduce a risk, possibly by making an attack less onerous, and organizations can choose to accept

---

[69] See, for example, Greenfield and Camm (2005), AFPAM 90-803 (2017), and Department of the Army Pamphlet 385-30 (2014).

[70] We recognize that *control* has a specific meaning in the cybersecurity community, as in AFPAM 90-803 (2017), but we are using the word as it used more generally across communities.

some risk, which paves the way to considering options for increasing resilience to, responding to, or recovering from an attack.

Figure E.1. Five-Step Risk Management Process

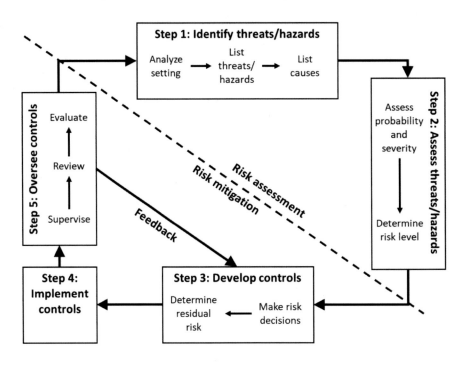

SOURCE: Authors' adaptation of Greenfield and Camm (2005), based on U.S. military guidance.

Thus, mitigation in this context includes responding, recovering, and building resilience (see Appendix A on vocabulary), which are familiar concepts in the cyber domain and are represented in NIST's cybersecurity framework (see the discussion later, under "NIST's Cybersecurity Functions"). In the five-step process, assessment informs mitigation, but its absence can also constrain mitigation. Looking for a pragmatic path to reconciling this tension, we consider, in Chapter 6, opportunities for leveraging the sequential process iteratively and continuously.

In this report, we have adopted the convention that risk is a combination of the *probability* of an event, brought on by a threat or hazard, and its *severity* in relation to its potential outcomes and their impacts or damage.[71] (See Figure E.2.) In the realm of cyber-related risks, probability consists of both the probability of a cyberattack attempt, which might depend partly on the relative attractiveness and expected impact of the attack, and the probability of the success of that attempt, if it is made. Within cybersecurity fields, this is commonly referred to as the "capability, intent, and access" of the threat actor, whom we refer to as the *attacker* (see

---

[71] For similar treatments, see Greenfield and Camm (2005), AFPAM (2017), and Department of the Army Pamphlet 385-30 (2014).

Appendix A on vocabulary). Severity, in turn, concerns the impact—or damage—associated with the set of possible outcomes of the attack, such as a loss of operability or a reduction in mission capability associated with a work stoppage.[72] This last point, that severity matters in relation to outcomes and impacts, not attacks per se, merits careful consideration because it plays a central role in establishing priorities.

Figure E.2. Simplified Risk Matrix

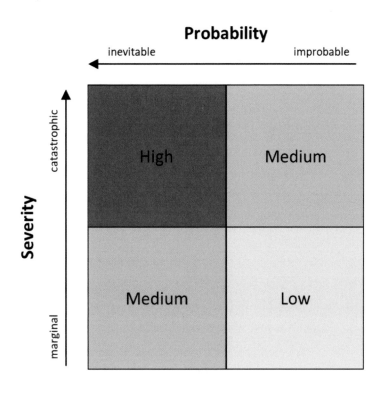

SOURCES: National Academies of Sciences, Engineering, and Medicine (2018, p. 148), and authors' adaptation of Greenfield and Camm (2005), based on U.S. military guidance.
NOTE: High (red), Medium (orange), and Low (yellow) signify risk levels and possible priorities.

The risk matrix, which places probability on one axis and severity on the other, suggests a first step toward prioritizing risk. For example, a risk that rates "High" might merit more immediate attention than one that rates "Low." That said, the assessment should be tethered to a benefit-cost analysis before acting on any particular hazard or threat (see, e.g., related discussions in U.S. Department of the Air Force, 2013; Department of the Army Pamphlet 385-30, 2014; and Greenfield and Paoli, 2013).

---

[72] As a practical matter, it might be difficult or impossible to trace an event to its ultimate impact, in which case it might be necessary to consider a proxy, such as the time value of delays or property loss.

We treat the risk matrix as a heuristic device, not as a strict analytical tool, because it obscures relevant complexity, uncertainty, and interdependencies between probability and severity, such as when the anticipated severity of an event affects its probability. A complete risk assessment would trace each possible attack from the probability of attempt, to the probability of success, to potential outcomes, and to the damage associated with the outcomes, but the range of possibilities at each stage is immense. (See, e.g., Ettinger, 2019, p. 111, for a graphic depiction that spans multiple dimensions.) Moreover, probability is not an independent point estimate but rather a potentially unknowable distribution that likely depends partly on expected impact, hence severity.[73]

Boolean Attack Model

Snyder et al. (2020, pp. 11–21) offers a complementary approach to conceptualizing risk that distinguishes between attackers' and defenders' perspectives and sheds light on the relationship between probability and severity by mapping an attack path for a cyber event with Boolean *and* and *or* statements (Figure E.3). Snyder et al. used the Boolean structure to assess cybersecurity and resilience in the context of cyber-related risks to weapon systems and missions in which a supply chain serves as a point of access, not as a target per se. Along the attack path, an attacker must have sufficient access, knowledge, *and* capability to be able to attack *and* anticipate sufficient impact to choose to attack, but none of these needs is static. Access, for example, is not once and for all but must be maintained over the relevant period of attack, which can extend over weeks, months, or years, and possibly defy detection. Moreover, by extension, to attack successfully, the attacker must not just anticipate impact but eventually obtain impact. The defender's problem is the mirror image of the attacker's, with *or* statements replacing *and* statements. It, too, would weigh the costs of acting against the benefits, i.e., impact avoidance.

In a more formal behavioral model (such as those found in game theory), the attacker might choose to incur the costs of obtaining access, knowledge, and capability if it *believed* the benefits of obtaining the impact would outweigh the costs. On that basis, a defender that can raise the costs of an attacker's access, knowledge, *or* capability *or* that can reduce the *expected* impact, hence benefits, of attacking, might deter an attacker.

However, an attacker might have varied means of obtaining access, knowledge, or capability, implying myriad *or* statements for each type of access, knowledge, and capability, such that a successful defender would need to block all means of obtaining one or the other, implying an equal number of subsidiary *and* statements. For example, an attacker might be able to obtain access through a back door *or* by phishing, which would obligate a defender to impede access through a back door *and* by phishing. Thus, if a defender focuses on one area, such as access, it

---

[73] For discussions of the analytical shortcomings of the matrix and concerns about its use, see, for example, National Academies of Sciences, Engineering, Engineering, and Medicine (2018), Cox (2008), and Rozell (2015).

cannot afford to miss anything because the attacker needs just one viable inroad among many possible options.

Figure E.3. Boolean Logic in Relation to Benefit-Cost Analysis

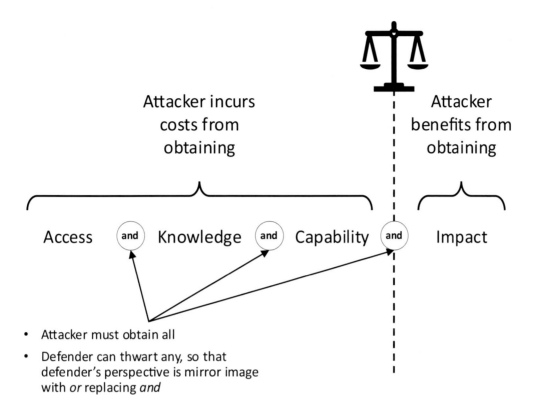

SOURCE: Authors' adaptation of figure in Snyder et al., 2020.

In terms of the risk matrix, access, knowledge, capability, and expected impact can all affect probability, but severity manifests in actual impact. Linking the two approaches shows that severity and probability are interrelated; an attacker's decision about whether to attack might be predicated on the expected impact of the attack and, hence, on its anticipated severity. While analytically complex, this relationship points to a potential opportunity for mitigation that we explore elsewhere in this report: Taking actions to reduce impact—or improve resilience—can reduce risk both directly and indirectly. Measures to improve resilience can both reduce the impact and, hence, the severity of an attack and, if they affect an adversary's expectations, reduce the probability of attack.

NIST's Cybersecurity Functions

In this section, we address the relationship between the five-step risk-management process and the cybersecurity functions (identify, protect, detect, respond, and recover) set out in NIST's cybersecurity framework.[74] The risk management process, especially the delineations of risk assessment and mitigation activities, is a common lens through which DoD conceptualizes risk and designs policies and measures to address it. The NIST functions instead suggest specific needs and opportunities for cybersecurity that relate to those activities. Although the functions were written for the cybersecurity of individual organizations involved in critical infrastructure rather than for complex networks of organizations that make up supply chains, NIST claims the potential for other, farther-reaching applications:

> While this document was developed to improve cybersecurity risk management in critical infrastructure, the Framework can be used by organizations in any sector or community. The Framework enables organizations—regardless of size, degree of cybersecurity risk, or cybersecurity sophistication—to apply the principles and best practices of risk management to improving security and resilience. (NIST, 2018b, p. v)

Extending the claim of breadth, we bring these functions to bear on cyber SCRM in this report by taking up the general principles of security and resilience. NIST's cybersecurity functions— identify, protect, detect, respond, and recover (Box E.1)—form the core of its framework and, as we demonstrate in Figure E.4, span risk assessment and mitigation (NIST, 2018b, pp. 7–8).

Figure E.4 explicitly links the NIST functions to the five-step risk management process and to risk itself. Response and recovery speak most directly to reducing the severity or impact of an event, whether an attack is directed through or on a supply chain. However, activities undertaken in the context of identification, protection, and detection can also affect impact, as a matter of resilience. Insomuch as fostering resilience ex ante—before the initiation of an attack—would reduce impact and, thus, lessen the need for response and recovery ex post—after the attack—we can think of resilience as a preemptive contributor to response and recovery. While NIST does not call out resilience in a separate function, it cites resilience as an end goal and addresses resilience in the context of activities for identification, protection, and recovery (NIST, 2018b, p. v). Detection resides at the cusp of ex ante and ex post approaches to risk reduction and might be considered a means of prevention insomuch as it can preempt an attack taking root.

---

[74] According to NIST (2020b), "The Cybersecurity Framework is a voluntary framework for reducing cyber risks to critical infrastructure. It is based on existing standards, guidelines, and practices, and was originally developed with stakeholders in response to Executive Order (EO) 13636 (February 12, 2013)." For information on NIST's related work on cybersecurity and SCRM, see NIST (2020a).

Box E.1. NIST's Cybersecurity Functions

NIST (2018b, pp. 7–8) defines the terms *identify, protect, detect, respond*, and *recover* as follows, along with sets of categories of outcomes and activities to flesh each out:

Identify. "Develop an organizational understanding to manage cybersecurity risk to systems, people, assets, data, and capabilities." The categories of outcomes for this function, which includes activities that are foundational for the effective use of the framework, consist of asset management, business environment, governance, risk assessment, risk management strategy, and supply chain risk management.

Protect. "Develop and implement appropriate safeguards to ensure delivery of critical services." The categories of outcomes for this function, which is intended to support an organization's ability to limit or contain the impact of a potential cybersecurity incident, consist of identity management, authentication and access control, awareness and training, data security, information protection processes and procedures, maintenance, and protective technology.

Detect. "Develop and implement appropriate activities to identify the occurrence of a cybersecurity event." The categories of outcomes for this function, which is intended to enable the timely discovery of an incident, consist of anomalies and events, continuous monitoring for security, and detection processes.

Respond. "Develop and implement appropriate activities to take action regarding a detected cybersecurity incident." The outcome categories for this function, which is intended to support the ability to contain the impact of an incident, consist of response planning execution and maintenance, communications (including voluntary external information-sharing), analysis, mitigation, and improvements in response activities. Mitigation, in this context, covers activities to prevent the expansion of an event, mitigate its effects, and resolve the incident.

Recover. "Develop and implement appropriate activities to maintain plans for resilience and to restore any capabilities or services that were impaired due to a cybersecurity incident." The outcome categories for this function, which is intended to support timely recovery to normal operations to reduce the impact of an event, consist of recovery planning execution and maintenance, improvements in recovery planning and processes, and communications.

Figure E.4. Comparison of Risk Assessment and Mitigation in the Five-Step Risk Management Process and NIST's Cybersecurity Functions

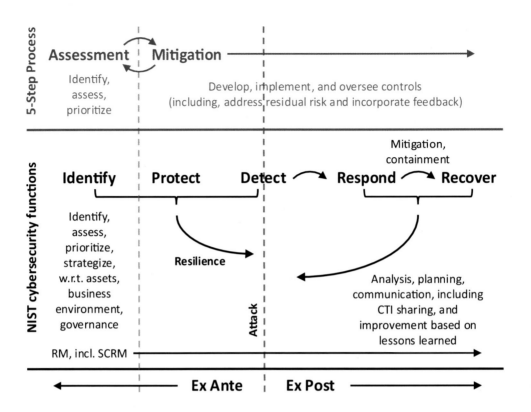

NOTE: In the risk management process, prioritization occurs in Step 2, "assess threats or hazards," which includes "evaluate probability and severity" and "determine risk level." RM = risk management; CTI = cyber threat intelligence; w.r.t. = with respect to.

Abbreviations

AFPAM	Air Force Pamphlet
AFRL	Air Force Research Laboratory
CMMC	Cybersecurity Maturity Model Certification
CSRC	NIST Computer Security Resource Center
DAF	Department of the Air Force
DoD	U.S. Department of Defense
EO	Executive Order
IO	input-output
IT	information technology
NDIA	National Defense Industrial Association
NIST	National Institute of Standards and Technology
OECD	Organisation for Economic Co-operation and Development
OT	operating technology
SCRM	supply chain risk management
TTX	tabletop exercise

References

Ablon, Lillian, Martin C. Libicki, and Andrea M. Abler, *Markets for Cybercrime Tools and Stolen Data: Hackers' Bazaar*, RAND Corporation, RR-610-JNI, 2014. As of July 25, 2023: https://www.rand.org/pubs/research_reports/RR610.html

Acemoglu, Daron, Vasco M. Carvalho, Asuman Ozdaglar, and Alireza Tahbaz-Salehi, "The Network Origins of Aggregate Fluctuations," *Econometrica*, Vol. 80, No. 5, September 2012.

Acemoglu, Daron, Asuman Ozdaglar, and Alireza Tahbaz-Salehi, "Systemic Risk and Stability in Financial Networks," *American Economic Review*, Vol. 105, No. 2, February 2015.

AFPAM—*See* Air Force Pamphlet.

Air Force Instruction 17-101, *Risk Management Framework (RMF) for Air Force Information Technology (IT)*, U.S. Department of the Air Force, February 6, 2020.

Air Force Pamphlet 90-803, *Risk Management (RM) Guidelines and Tools*, U.S. Department of the Air Force, 2013, current as of March 3, 2017. As of July 24, 2023: https://static.e-publishing.af.mil/production/1/af_se/publication/dafpam90-803/dafpam90-803.pdf

Aldasoro, Iñaki, Leonardo Gambacorta, Paolo Giudici, and Thomas Leach, *The Drivers of Cyber Risk*, Bank for International Settlements, May 2020. As of July 24, 2023: https://www.bis.org/publ/work865.pdf

Aven, Terje, Yakov Ben-Haim, Henning Boje Andersen, Tony Cox, Enrique López Droguett, Michael Greenberg, Seth Guikema, Wolfgang Kröger, Ortwin Renn, Kimberly M. Thompson, and Enrico Zio, "Society for Risk Analysis Glossary," Society for Risk Analysis, August 2018.

Baghalian, Atefeh, Shabnam Rezapour, and Reza Zanjirani Farahani, "Robust Supply Chain Network Design with Service Level Against Disruptions and Demand Uncertainties: A Real-Life Case," *European Journal of Operational Research*, Vol. 227, No. 1, 2013.

Baliga, Sandeep, Ethan Bueno de Mesquita, and Alexander Wolitzky, "Deterrence with Imperfect Attribution," *American Political Science Review*, Vol. 114, No. 4, November 2020.

Bandyopadhyay, Tridib, Varghese Jacob, and Srinivasan Raghunathan, "Information Security in Networked Supply Chains: Impact of Network Vulnerability and Supply Chain Integration on Incentives to Invest," *Information Technology and Management*, Vol. 11, No. 1, March 2010.

Bandyopadhyay, Tridib, Vijay S. Mookerjee, and Ram C. Rao, "Why IT Managers Don't Go for Cyber-Insurance Products," *Communications of the ACM*, Vol. 52, No. 11, November 2009.

Barrot, Jean-Noël, and Julien Sauvagnat, "Input Specificity and the Propagation of Idiosyncratic Shocks in Production Networks," *Quarterly Journal of Economics*, Vol. 131, No. 3, August 2016.

Bartock, Michael, Jeffrey Cichonski, Murugiah Souppaya, Matthew Smith, Greg Witte, and Karen Scarfone, *Guide for Cybersecurity Event Recovery*, National Institute of Standards and Technology, Special Publication 800-184, December 2016. As of July 2023: https://csrc.nist.gov/pubs/sp/800/184/final

Bell, M. G. H., U. Kanturska, J.-D. Schmöcker, and A. Fonzone, "Attacker-Defender Models and Road Network Vulnerability," *Philosophical Transactions of the Royal Society A*, Vol. 366, No. 1872, June 2008.

Bier, Vicki M., "Choosing What to Protect," *Risk Analysis*, Vol. 27, No. 3, June 2007.

Bier, Vicki, Santiago Oliveros, and Larry Samuelson, "Choosing What to Protect: Strategic Defensive Allocation Against an Unknown Attacker," *Journal of Public Economic Theory*, Vol. 9, No. 4, August 2007.

Bing, Chris, "You Can Now Buy a Mirai-Powered Botnet on the Dark Web," Cyberscoop, October 27, 2016.

Blosfield, Elizabeth, "Data Deficit Remains Key Challenge for Cyber Insurance Underwriters," *Insurance Journal*, June 18, 2019.

BlueVoyant, *Supply Chain Cyber Risk: Managing Cyber Risk Across the Extended Vendor Ecosystem*, 2020. As of July 25, 2023: https://f.hubspotusercontent10.net/hubfs/4896063/USA%20REPORT.pdf

Boehm, Christoph E., Aaron Flaaen, and Nitya Pandalai-Nayar, "Input Linkages and the Transmission of Shocks: Firm-Level Evidence from the 2011 Tōhoku Earthquake," *Review of Economics and Statistics*, Vol. 101, No. 1, March 2019.

Boyens, Jon, Angela Smith, Nadya Bartol, Kris Winkler, Alex Holbrook, and Matthew Fallon, *Cybersecurity Supply Chain Risk Management Practices for Systems and Organizations*, rev. 1, draft, National Institute of Standards and Technology, Special Publication 800-161, 2015.

Boyson, Sandor, "Cyber Supply Chain Risk Management: Revolutionizing the Strategic Control of Critical IT Systems," *Technovation*, Vol. 34, No. 7, July 2014.

Brady, Ryan R., and Victoria A. Greenfield, "Competing Explanations of U.S. Defense Industry Consolidation in the 1990s and Their Policy Implications," *Contemporary Economic Policy*, Vol. 28, No. 2, April 2010.

Carter, Ashton B., "Defense Industrial Base Cyber Security," memorandum, U.S. Department of Defense, October 31, 2012. As of July 2023:
https://www.acq.osd.mil/dpap/policy/policyvault/OSD012537-12-RES.pdf

Carvalho, Vasco M., Makoto Nirei, Yukiko U. Saito, and Alireza Tahbaz-Salehi, "Supply Chain Disruptions: Evidence from the Great East Japan Earthquake," *SSRN Electronic Journal*, 2016-01-01, 2020.

Carvalho, Vasco M., and Alireza Tahbaz-Salehi, "Production Networks: A Primer," *Annual Review of Economics*, Vol. 11, 2019.

Caton, Jeffrey L., *Army Support of Military Cyberspace Operations: Joint Contexts and Global Escalation Implications*, Army War College, 2015.

Center for Strategic and International Studies, "Significant Cyber Incidents," webpage, undated. As of December 20, 2020:
https://www.csis.org/programs/strategic-technologies-program/significant-cyber-incidents

Collins, Keith, "Inside the Digital Heist That Terrorized the World—and Only Made $100k," Quartz, May 21, 2017.

Cox, Louis Anthony, "What's Wrong with Risk Matrices?" *Risk Analysis*, Vol. 28, No. 2, April 2008.

Crosignani, Matteo, Marco Macchiavelli, and André F. Silva, "Pirates Without Borders: The Propagation of Cyberattacks Through Firms' Supply Chains," Federal Reserve Bank of New York, No. 937, July 2021.

Cybersecurity and Infrastructure Security Agency, *Assessment of the Cyber Insurance Market*, U.S. Department of Homeland Security, December 21, 2018. As of July 2023:
https://www.cisa.gov/sites/default/files/publications/
20_0210_cisa_oce_cyber_insurance_market_assessment.pdf

Cyberspace Solarium Commission, *United States of America Cyberspace Solarium Commission Report*, March 2020.

DAF—*See* Department of the Air Force.

Davis, John S., II, Benjamin Boudreaux, Jonathan William Welburn, Jair Aguirre, Cordaye Ogletree, Geoffrey McGovern, and Michael S.Chase, *Stateless Attribution: Toward International Accountability in Cyberspace*," RAND Corporation, RR-2081-MS, 2017. As of July 25, 2023:
https://www.rand.org/pubs/research_reports/RR2081.html

Defense Science Board, *Task Force on Cyber Supply Chain*, final report, Department of Defense, 2017.

Deloitte, *Black-Market Ecosystem: Estimating the Cost of "Pwnership,"* December 2018.

Department of the Army Pamphlet 385-30, *Risk Management*, December 2, 2014. As of July 24, 2023:
https://armypubs.army.mil/epubs/DR_pubs/DR_a/pdf/web/p385_30.pdf

Do, Cuong T., Nguyen H. Tran, Choongseon Hong, Charles A. Kamhoua, Kevin A. Kwiat, Erik Blasch, Shaolei Ren, Niki Pissinou, and Sundaraja Sitharama Iyengar, "Game Theory for Cyber Security and Privacy," *ACM Computing Surveys*, Vol. 50, No. 2, 2017.

DoD—*See* U.S. Department of Defense.

Dolgui, Alexandre, Dmitry Ivanov, and Boris Sokolov, "Ripple Effect in the Supply Chain: An Analysis and Recent Literature," *International Journal of Production Research*, Vol. 56, Nos. 1–2, 2018.

Dresher, Melvin, *Games of Strategy: Theory and Applications*, reprint, RAND Corporation, CB-149-1, [1961] 2007. As of July 27, 2023:
https://www.rand.org/pubs/commercial_books/CB149-1.html

Edwards, Benjamin, Alexander Furnas, Stephanie Forrest, and Robert Axelrod, "Strategic Aspects of Cyberattack, Attribution, and Blame," *Proceedings of the National Academy of Sciences*, Vol. 114, No. 11, 2017.

Ettinger, Jared, *Cyber Intelligence Tradecraft Report: The State of Cyber Intelligence Practices in the United States*, Carnegie Mellon Software Engineering Institute, 2019.

Executive Order 14017, "America's Supply Chains," Executive Office of the President, February 24, 2021.

Executive Order 14028, "Improving the Nation's Cybersecurity," Executive Office of the President, May 12, 2021.

Fiksel, Joseph, "From Risk to Resilience," in Joseph Fiksel, *Resilient by Design: Creating Businesses That Adapt and Flourish in a Changing World*, Springer, 2015.

Gonzales, Daniel, Sarah Harting, Mary Kate Adgie, Julia Brackup, Lindsey Polley, and Karlyn D. Stanley, *Unclassified and Secure: A Defense Industrial Base Cyber Protection Program for Unclassified Defense Networks*, RAND Corporation, RR-4227-RC, 2020. As of July 27, 2023:
https://www.rand.org/pubs/research_reports/RR4227.html

Greenberg, Andy, *Sandworm: A New Era of Cyberwar and the Hunt for the Kremlin's Most Dangerous Hackers*, Anchor, 2020.

Greenfield, Victoria A., and Frank Camm, *Risk Management and Performance in the Balkans Support Contract*, RAND Corporation, MG-282-A, 2005. As of July 27, 2023: https://www.rand.org/pubs/monographs/MG282.html

Greenfield, Victoria A., and Letizia Paoli, "A Framework to Assess the Harms of Crimes," *British Journal of Criminology*, Vol. 53, No. 5, September 2013.

Greenfield, Victoria A., and David M. Persselin, *An Economic Framework for Evaluating Military Aircraft Replacement*, RAND Corporation, MR 1489-AF, 2002. As of July 27, 2023: https://www.rand.org/pubs/monograph_reports/MR1489.html

Greenfield, Victoria A., and David M. Persselin, "How Old Is Too Old? An Economic Approach to Replacing Military Aircraft," *Defence and Peace Economics*, Vol. 14, No. 5, October 2003.

Haphuriwat, Naraphorn, Vicki M. Bier, and Henry H. Willis, "Deterring the Smuggling of Nuclear Weapons in Container Freight Through Detection and Retaliation," *Decision Analysis*, Vol. 8, No. 2, June 2011.

Huang, Keman, Michael Siegel, and Stuart Madnick, "Systematically Understanding the Cyber Attack Business: A Survey," *ACM Computing Surveys*, Vol. 51, No. 4, July 2018.

Hausken, Kjell, "Information Sharing Among Firms and Cyber Attacks," *Journal of Accounting and Public Policy*, Vol. 26, No. 6, November–December 2007.

Hausken, Kjell, "Strategic Defense and Attack for Series and Parallel Reliability Systems," *European Journal of Operational Research*, Vol. 186, No. 2, April 2008.

Herr, Trey, "Governing Proliferation in Cybersecurity," *Global Summitry*, Vol. 3, No. 1, 2017.

Huijgens, Hennie, Arie van Deursen, Leandro L. Minku, and Chris Lokan, "Effort and Cost in Software Engineering: A Comparison of Two Industrial Data Sets," *Proceedings of the 21st International Conference on Evaluation and Assessment in Software Engineering*, Association for Computing Machinery, June 2017.

Jackson, Brian A., and Tom LaTourrette, "Assessing the Effectiveness of Layered Security for Protecting the Aviation System Against Adaptive Adversaries," *Journal of Air Transport Management*, Vol. 48, September 2015.

Joint Publication 3-12, *Cyberspace Operations*, Joint Chiefs of Staff, June 8, 2018.

Kamalahmadi, Masoud, and Mahour Mellat Parast, "A Review of the Literature on the Principles of Enterprise and Supply Chain Resilience: Major Findings and Directions for Future Research," *International Journal of Production Economics*, Vol. 171, No. 1, January 2016.

Kessel, Frank, ed., *Psychology, Science, and Human Affairs: Essays in Honor of William Bevan*, Routledge, 1995.

Kunreuther, Howard, and Geoffrey Heal, "Interdependent Security," *Journal of Risk and Uncertainty*, Vol. 26, Nos. 2–3, March/May 2003.

Lempert, Robert J., David G. Groves, Steven W. Popper, and Steve C. Bankes, "A General, Analytic Method for Generating Robust Strategies and Narrative Scenarios," *Management Science*, Vol. 52, No. 4, April 2006.

Leontief, Wassily, *Input-Output Economics*, Oxford University Press, 1966.

Libicki, Martin C., *Cyberdeterrence and Cyberwar*, RAND Corporation, MG-877-AF, 2009. As of July 27, 2023:
https://www.rand.org/pubs/monographs/MG877.html

Lorell, Mark A., Julia Lowell, Richard M. Moore, Victoria Greenfield, and Katia Vlachos, *Going Global? U.S. Government Policy and the Defense Aerospace Industry*, RAND Corporation, MR-1537-AF, 2002. As of November 6, 2023:
https://www.rand.org/pubs/monograph_reports/MR1537.html

MacKenzie, Cameron A., Kash Barker, and Joost R. Santos, "Modeling a Severe Supply Chain Disruption and Post-Disaster Decision Making with Application to the Japanese Earthquake and Tsunami," *IIE Transactions*, Vol. 46, No. 12, 2014.

Marotta, Angelica, Fabio Martinelli, Stefano Nanni, Albina Orlando, and Artsiom Yautsiukhin, "Cyber-Insurance Survey," *Computer Science Review*, Vol. 24, May 2017.

Mathews, Lee, "World's Biggest Mirai Botnet Is Being Rented Out for DDoS Attacks," *Forbes*, November 29, 2016.

Moore, Nancy Y., Elvira N. Loredo, Amy G. Cox, and Clifford A. Grammich, *Identifying and Managing Acquisition and Sustainment Supply Chain Risks*, RAND Corporation, RR-549-AF, 2015. As of July 27, 2023:
https://www.rand.org/pubs/research_reports/RR549.html

Morral, Andrew R., and Brian A. Jackson, *Understanding the Role of Deterrence in Counterterrorism Security*, RAND Corporation, OP-281-RC, 2009. As of July 27, 2023:
https://www.rand.org/pubs/occasional_papers/OP281.html

Nagurney, Anna, Ladimer S. Nagurney, and Shivani Shukla, "A Supply Chain Game Theory Framework for Cybersecurity Investments Under Network Vulnerability," in Nicholas J. Daras and Michael Th. Rassias, eds., *Computation, Cryptography, and Network Security*, Springer International Publishing, 2015.

Nagurney, Anna, and Shivani Shukla, "Multifirm Models of Cybersecurity Investment Competition vs. Cooperation and Network Vulnerability," *European Journal of Operational Research*, Vol. 260, No. 2, July 2017.

Nash, Kim S., Sara Castellanos, and Adam Janofsky, "One Year After NotPetya Cyberattack, Firms Wrestle with Recovery Costs," *Wall Street Journal*, June 27, 2018.

National Academies of Sciences, Engineering, Engineering, and Medicine, *Reducing the Threat of Improvised Explosive Device Attacks by Restricting Access to Explosive Precursor Chemicals*, National Academies Press, 2018.

National Defense Industrial Association, *Cybersecurity for Manufacturing Networks*, Cybersecurity for Advanced Manufacturing Joint Working Group, October 2017. As of July 2023:
https://www.ndia.org/-/media/sites/ndia/divisions/working-groups/cfam/ndia-cfam-2017-white-paper-20171023.ashx?la=en

National Institute of Standards and Technology, "The Five Functions," webpage, 2018a. As of July 27, 2023:
https://www.nist.gov/cyberframework/online-learning/five-functions

National Institute of Standards and Technology, *Framework for Improving Critical Infrastructure Cybersecurity*, Version 1.1, April 16, 2018b.

National Institute of Standards and Technology, "Cyber Supply Chain Risk Management C-SCRM," webpage, 2020a. As of July 27, 2023:
https://csrc.nist.gov/Projects/cyber-supply-chain-risk-management

National Institute of Standards and Technology, "Cybersecurity Framework," webpage, 2020b. As of December 2020:
https://csrc.nist.gov/Topics/Applications/cybersecurity-framework

National Institute of Standards and Technology Computer Security Resource Center, "Glossary," webpage, undated. As of July 24, 2023:
https://csrc.nist.gov/glossary

NDIA—*See* National Defense Industrial Association.

Nissen, Chris, John Gronager, Robert Metzger, and Harvey Rishikof, *Deliver Uncompromised: A Strategy for Supply Chain Security and Resilience in Response to the Changing Character of War*, MITRE Corporation, 2018.

NIST—*See* National Institute of Standards and Technology.

O'Connell, Caolionn, Elizabeth Hastings Roer, Rick Eden, Spencer Pfeifer, Yuliya Shokh, Lauren A. Mayer, Jake McKeon, Jared Mondschein, Phillip Carter, Victoria A. Greenfield, and Mark Ashby, *Managing Risk in Globalized Supply Chains*, RAND Corporation, RR-A425-1, 2021. As of July 27, 2023:
https://www.rand.org/pubs/research_reports/RRA425-1.html

OECD—*See* Organisation for Economic Co-operation and Development.

Office of the Under Secretary of Defense for Acquisition & Sustainment, "Cybersecurity Maturity Model Certification," webpage, undated. As of December 20, 2020:
https://web.archive.org/web/20201221160658/https://www.acq.osd.mil/cmmc/

Organisation for Economic Co-operation and Development, *Supporting an Effective Cyber Insurance Market: OECD Report for the G7 Presidency*, May 2017.

Paoli, Letizia, and Victoria A. Greenfield, "Starting from the End: A Plea for Focusing on the Consequences of Crime," *European Journal of Crime, Criminal Law and Criminal Justice*, Vol. 23, No. 2, April 18, 2015.

Pomerleau, Mark, "Here's How DoD Organizes Its Cyber Warriors," C4ISRNET, July 25, 2017. As of July 27, 2023:
https://www.c4isrnet.com/workforce/career/2017/07/25/heres-how-dod-organizes-its-cyber-warriors/

Presidential Policy Directive 21, "Critical Infrastructure Security and Resilience," Executive Office of President Barack Obama, February 12, 2013.

Romanosky, Sasha, Lillian Ablon, Andreas Kuehn, and Therese Jones, "Content Analysis of Cyber Insurance Policies: How Do Carriers Price Cyber Risk?" *Journal of Cybersecurity*, Vol. 5, No. 1, 2019.

Romanosky, Sasha, and Benjamin Boudreaux, "Private-Sector Attribution of Cyber Incidents: Benefits and Risks to the U.S. Government," *International Journal of Intelligence and CounterIntelligence*, Vol. 34, No. 3, 2021.

Roy, Sankardas, Charles Ellis, Sajjan Shiva, Dipankar Dasgupta, Vivek Shandilya, and Qishi Wu, "A Survey of Game Theory as Applied to Network Security," 43rd Hawaii International Conference on System Sciences, 2010.

Rozell, Daniel J., "A Cautionary Note on Qualitative Risk Ranking of Homeland Security Threats," *Homeland Security Affairs*, Vol. 11, No. 3, February 2015.

Santos, Joost R., and Yacov Y. Haimes, "Modeling the Demand Reduction Input-Output (I-O) Inoperability Due to Terrorism of Interconnected Infrastructures," *Risk Analysis*, Vol. 24, No. 6, December 2004.

Santos, Joost R., Yacov Y. Haimes, and Chenyang Lian, "A Framework for Linking Cybersecurity Metrics to the Modeling of Macroeconomic Interdependencies," *Risk Analysis*, Vol. 27, No. 5, October 2007.

Schelling, Thomas C., *The Strategy of Conflict*, Harvard University Press, 1960.

Simon, Jay, and Ayman Omar, "Cybersecurity Investments in the Supply Chain: Coordination and a Strategic Attacker," *European Journal of Operational Research*, Vol. 282, No. 1, April 2020.

Snyder, Don, Lauren A. Mayer, Guy Weichenberg, Danielle C. Tarraf, Bernard Fox, Myron Hura, Suzanne Genc, and Jonathan William Welburn, *Measuring Cybersecurity and Cyber Resiliency*, RAND Corporation, RR-2703-AF, 2020. As of July 27, 2023: https://www.rand.org/pubs/research_reports/RR2703.html

Snyder, Don, James D. Powers, Elizabeth Bodine-Baron, Bernard Fox, Lauren Kendrick, and Michael H. Powell, *Improving the Cybersecurity of U.S. Air Force Military Systems Throughout Their Life Cycles*, RAND Corporation, RR-1007-AF, 2015. As of July 27, 2023: https://www.rand.org/pubs/research_reports/RR1007.html

Tajitsu, Naomi, "Five Years After Japan Quake, Rewiring of Auto Supply Chain Hits Limits," Reuters, March 29, 2016.

Tajitsu, Naomi, and Makiko Yamazaki, "Toyota, Other Major Japanese Firms Hit by Quake Damage, Supply Disruptions," Reuters, April 17, 2016.

Thomson, Iain, "NotPetya Ransomware Attack Cost Us $300m—Shipping Giant Maersk," online article, *The Register*, August 16, 2017.

U.S. Cyber Command, "All Cyber Mission Force Teams Achieve Initial Operating Capability," press release, U.S. Department of Defense, October 24, 2016.

U.S. Cyber Command, "Cyber Mission Force Achieves Full Operational Capability," press release, U.S. Department of Defense, May 17, 2018.

U.S. Department of Defense, *Assessing and Strengthening the Manufacturing and Defense Industrial Base and Supply Chain Resiliency of the United States*, September 2018.

U.S. Government Accountability Office, "Federal Agencies Need to Take Urgent Action to Manage Supply Chain Risks," GAO-21-171, December 2020.

Wagner, Stefan, and Melanie Ruhe, "A Systematic Review of Productivity Factors in Software Development," arXiv preprint arXiv:1801.06475, January 19, 2018.

Wang, Shaun S., "Integrated Framework for Information Security Investment and Cyber Insurance," *Pacific-Basin Finance Journal*, Vol. 57, October 2019.

Welburn, Jonathan, Justin Grana, and Karen Schwindt, "Cyber Deterrence with Imperfect Attribution and Unverifiable Signaling," *European Journal of Operational Research*, Vol. 306, No. 3, May 2023.

Welburn, Jonathan William, and Aaron Strong, "Systemic Cyber Risk and Aggregate Impacts," *Risk Analysis*, Vol. 42, No. 8, August 2022.

Welburn, Jonathan William, Aaron Strong, Florentine Eloundou Nekoul, Justin Grana, Krystyna Marcinek, Osonde A. Osoba, Nirabh Koirala, and Claude Messan Setodji, *Systemic Risk in the Broad Economy: Interfirm Networks and Shocks in the U.S. Economy*, RAND Corporation, RR-4185-RC, 2020. As of March 4, 2022:
https://www.rand.org/pubs/research_reports/RR4185.html

White House, *Building Resilient Supply Chains, Revitalizing American Manufacturing, and Fostering Broad-Based Growth: 100-Day Reviews Under Executive Order 14017*, June 2021.

Zhuang, Jun, Vicki M. Bier, and Oguzhan Alagoz, "Modeling Secrecy and Deception in a Multiple-Period Attacker–Defender Signaling Game," *European Journal of Operational Research*, Vol. 203, No. 2, June 2010.